VEHICLE EXTRICATION
A PRACTICAL GUIDE

VEHICLE EXTRICATION
A PRACTICAL GUIDE

Fire Engineering

BRIAN G. ANDERSON

> **DISCLAIMER**
>
> The recommendations, advice, descriptions, and methods in this book are presented solely for educational purposes. The author and publisher assume no liability whatsoever for any loss or damage that results from the use of any of the material in this book. Use of the material in this book is solely at the risk of the user.

Copyright ©2005 by
PennWell Corporation
1421 South Sheridan Road
Tulsa, Oklahoma 74112-6600 USA
800.752.9764
+1.918.831.9421
sales@pennwell.com
www.pennwellbooks.com
www.fireengineeringbooks.com
www.pennwell.com

Supervising Editor: Jared d'orr Wicklund
Production Editor: Sue Rhodes Dodd
Cover Design: Ken Wood
Book Design: Wes Rowell

Library of Congress Cataloging-in-Publication Data

Anderson, Brian G.
  Vehicle extrication : a practical guide / by Brian G. Anderson.-- 1st American ed.
    p. cm.
  ISBN 1-59370-021-0 (0)
  1. Traffic accidents. 2. Crash injuries. 3. Transport of sick and wounded. 4. Rescue work.
I. Title.
RC88.9.T7A535 2004
617.1'028--dc22

2004014882

All rights reserved. No part of this book may be reproduced, stored in a retrieval system, or transcribed in any form or by any means, electronic or mechanical, including photocopying and recording, without the prior written permission of the publisher.

Printed in the United States of America
1 2 3 4 5   09 08 07 06 05

For Georgia

Contents

Photo Credits xv

Acknowledgments xvii

Acronyms xxiii

1—Getting Organized 1
    Sharp Firefighters, Powerful Tools, and Strong Leadership 1
    Gaining Knowledge and Skill 2
    Role and Responsibility of the Company Officer 3
        Making tough decisions 3
        Size-up 3
        Presence of hazards 5
        Determining adequacy of resources 5
        Developing the tactical plan and assigning tasks 8
        The basic tactical options 11
        Risk/benefit analysis 13
        Plan A, B, and C 14
        Extrication time estimates 15
        Things the company officer shouldn't be doing 16
    Role and Responsibility of the Driver Operator 17
    Role and Responsibility of the Firefighter 18
    Summary 18
    Appendix: Resource Utilization 19
        Case Study: Overturned Tractor Trailer at the Edge of the Everglades 19

2—Tools, Equipment, and Apparatus 21
    Introduction 21
    Personal Protective Equipment 21
        Coats, pants, and boots 22
        Gloves 22
        Head, face, and eye protection 23
        The issue of compliance 24
    Tools & Equipment 24
        Cribbing 24
        Chains 26
        Chain construction and grading 28
        Other components of a chain set 28
        Typical chain configurations 30
        Come-alongs 32
        Come-along components 32
        Basic come-along actions 34

Air (pneumatic) tool systems	42
The air tool system overview	43
Air sources	43
Regulators	45
Air hoses	48
Hose couplings	50
Important points for departments converting to 4500-psi SCBAs	50
Setting up the system	51
Troubleshooting	51
Air Chisels	52
Impact wrenches	63
Whizzer/die grinder	63
Air bags	65
Electric Tools	67
Manual Hydraulic Tools	69
Bottle jacks	69
Basic jack operation—to raise	69
Basic jack operation—to lower	70
Floor jack	70
Basic floor jack operations—to raise	70
Basic floor jack operations—to lower	70
Manually operated spreaders	71
Automotive repair-style spreaders	71
Modern manually operated spreader tools	71
Basic spreader operations—to spread	72
Basic spreader operations—to close	72
Manually operated cutters	72
Basic operation—to cut/close	72
Basic operation—to open	72
Engine-Driven Hydraulic Tools—a Historical Perspective	72
Modern spreaders, cutters, and rams	76
Power plants—a closer look	77
Apparatus	85
Appendix: Apparatus Features and Tool Storage	86
3—Stabilization	**95**
Stabilizing the Scene	95
Protecting the scene with effective apparatus placement	99
Identifying utility hazards at the scene core	101
Vehicles in contact with live wires	103

Contents

Stabilization of the Vehicle 103
   Shutting down the vehicle 103
   The hybrids 105
   Fire control and extinguishment 107
   Vehicle stabilization—stopping large movements 108
   Vehicle stabilization—stopping small movements 108
Stabilization of the Victim 110
Summary 111
Procedures 113
   Procedure 1—Engine Compartment Access Using a Halligan Adz 114
   Procedure 2—Engine Compartment Access Using a Halligan Pike to Pry a Corner 115
   Procedure 3—Engine Compartment Access Using a Halligan Pike to Punch a Hole 116
   Procedure 4—Engine Compartment Access Using a Saw (Front Latch Cuts) 118
   Procedure 5—Engine Compartment Access Using a Saw (Center Cuts) 119
   Procedure 6—Engine Compartment Access Using a Saw (Straight Cuts) 120
   Procedure 7—Engine Compartment Access Using a Saw (Corner Cuts) 121
   Procedure 8—Using Cribbing to Stabilize a Vehicle on Its Side 122
   Procedure 9—Stabilizing a Vehicle to Prevent Further Forward Movement 124
   Procedure 10—Stabilizing a Vehicle on Its Side with a Come-along 127
   Procedure 11—Using a Long 4x4 to Stabilize a Vehicle on Its Side 129
   Procedure 12—Stabilizing a Vehicle from Both Sides
      Using a Capabear Stabilization Device 130
   Procedure 13—Stabilizing a Vehicle from One Side 132
   Procedure 14—Stabilizing a Vehicle on Its Roof with Step Chocks,
      Capabears, and Box Cribbing 136
   Procedure 15—Stabilization on the Roof 137
   Procedure 16—Stabilizing an Overturned Vehicle with Stump Screw Jacks 138
   Procedure 17—Stabilizing a Vehicle on a Jersey Barrier 140

4—Door and Side Procedures 141
Terminology of Components and Procedures 141
The Size-up and Action Plan 142
Selecting the Best Procedures and Tools 144
The Procedure Steps 144
Procedures 147
   Procedure 1—Removing Tempered Glass with a Centered Punch 148
   Procedure 2—Removing Tempered Glass with a Halligan 150
   Procedure 3—Creating a Purchase Point with a Halligan 151
   Procedure 4—The *Pinch and Curl* Purchase Point 152
   Procedure 5—Popping a Door Open with an Air Chisel 153

Procedure 6—Popping a Door Open with Hydraulic Spreaders 155
Procedure 7—Vertical Spread to Pop a Door Open 157
Procedure 8—Popping a Rear Door with Hydraulic Spreaders 160
Procedure 9—Manually Widening a Door Opening 162
Procedure 10—Widening a Door Opening with a Come-along 163
Procedure 11—Removing a Door with an Impact Wrench 166
Procedure 12—Removing a Door at the Hinges with an Air Chisel 167
Procedure 13—Removing a Door with Spreaders—Exterior Approach 171
Procedure 14—Removing a Door with Spreaders from the Interior 175
Procedure 15—Removing a Door from the Interior with Cutters 177
Procedure 16—Removing a Door from the Exterior with Cutters 179
Procedure 17—Popping and Removing a Van Door 181
Procedure 18—Creating a Third Door with an Air Chisel 183
Procedure 19—Creating a Third Door with Hydraulic Tools 188
Procedure 20—Creating a Fourth Door with an Air Chisel .192
Procedure 21—Side-Out 196
Procedure 22—Creating a Fourth Door with Hydraulic Tools 203
Procedure 23—Opening a Trunk with Hand Tools 206
Procedure 24—Opening a Trunk with Hydraulic Spreaders .208

5—**Roof Procedures** 209
  The Roof Structure 210
  A Practical Approach for Sizing Up the Roof 210
  Universal Precautions for Crew Safety When Performing Roof Procedures 211
  Existence and Non-Existence of Air Bag Labels 213
  Technology Advancements 213
  Gas Lifters 214
  Selecting the Best Tools for the Job 215
  Selecting the Best Procedure 216
  Summary 216
  Procedures 217
    Procedure 1—Quickly Establishing the Presence or Absence of Inflatable Curtains
      to Determine the Likelihood of the Presence of Compressed Gas Cylinders 218
    Procedure 2—Checking "A" and "B" Posts for Compressed Gas Cylinders
      in Vehicles Equipped with Inflatable Curtains 219
    Procedure 3—Checking the "C" Posts for Compressed Gas Cylinders
      in Vehicles Equipped with Inflatable Curtains 220
    Procedure 4—Checking the Roof Edge for Compressed Gas Cylinders
      in Vehicles Equipped with Inflatable Curtains 221
    Procedure 5—Lifting a Crushed Roof with a Hi-Lift Jack 222

Contents

    Procedure 6—Lifting a Crushed Roof with a Hydraulic Spreader 223
    Procedure 7—Lifting a Crushed Roof with a Hydraulic Ram 224
    Procedure 8—Removing Laminated Glass 226
    Procedure 9—Basic Roof Flapping Procedure—Why Striking the Roof
        is often Necessary 230
    Procedure 10—Flapping a Roof with Hand Tools 231
    Procedure 11—Flapping or Removing a Roof with an Air Chisel 234
    Procedure 12—Flapping a Roof with Hydraulic Tools 241
    Procedure 13—Flapping a Roof with a Come-along 244
    Procedure 14—Removing a Roof with an Electric Reciprocating Saw 248
    Procedure 15—Removing a Roof with Hydraulic Tools 251
    Procedure 16—Making a Three-sided Roof Cut with an Air Chisel 256
    Procedure 17—Making a Three-sided Roof Cut with a Reciprocating Saw 262
    Procedure 18—Removing the Center of the Roof with a Reciprocating Saw 267
    Procedure 19—Roof Flap with Hydraulic Cutters, Vehicle on Its Side 269

6—Interior Procedures 273
    Relocating a steering column with a hydraulic spreader or come-along 274
    Performing a dash roll-up with hydraulic tools 274
    Removing pedals 275
    Relocating a pedal with a length of rope 276
    Using a Hi-Lift jack to create interior space 276
    Relocating a seat 276
  Summary 277
  Procedures 279
    Procedure 1—Interior Spread with a Come-Along 280
    Procedure 2—Relocation of a Front Seat with a Come-Along .282
    Procedure 3—Interior Spread with a Winch 286
    Procedure 4—Moving a Seat Rearward with Spreaders 288
    Procedure 5—Removing a Front Seat with an Air Chisel 289
    Procedure 6—Removing a Seat Back with Hydraulic Cutters 292
    Procedure 7—Removing a Front Seat with Hydraulic Tools 293
    Procedure 8—Relocating a Steering Ring with Bolt Cutters 294
    Procedure 9—Relocating a Steering Column with a Come-Along 295
    Procedure 10—Relocating a Steering Column with a Hydraulic Spreader 299
    Procedure 11—Removing a Steering Wheel with Hydraulic Cutters 305
    Procedure 12—Relocating a Pedal with a Length of Rope 306
    Procedure 13—Relocating a Pedal Across the Hood with a Come-along 307
    Procedure 14—Relocating a Pedal through the Passenger's Door with a Come-along . . .309
    Procedure 15—Relocating a Pedal through the Rear Windshield with a Come-along . . .311

        Procedure 16—Relocating a Pedal over the Roof with a Come-along 313
        Procedure 17—Removing a Pedal with a Whizzer 315
        Procedure 18—Removing a Pedal with a Manually-operated Hydraulic Cutter 316
        Procedure 19—Removing a Pedal with Hydraulic Cutters 318
        Procedure 20—Using a Hi-Lift Jack to Create Interior Space 319
        Procedure 21—Interior Spread with a Hydraulic Ram 321
        Procedure 22—Special Interior Problems in Police Cars 323
        Procedure 23—Performing a Dash Roll with a Hi-Lift Jack 325
        Procedure 24—Dash Roll-up with Hydraulic Tools 326
        Procedure 25—Dash Lift 332

7—Crash Related Impalement 339
    Procedures 345
        Procedure 1—Cutting Rebar with Hydraulic Cutters 346
        Procedure 2—Heat-Fence—A Welder's Product that Reduces Heat Transfer 347
        Procedure 3—Cutting Rebar with a Reciprocating Saw 348
        Procedure 4—Cutting Large Pipes with a Reciprocating Saw 349
        Procedure 5—Cutting Large Pipes with a Whizzer 350
        Procedure 6—Cutting Large Pipes with a Hydraulic Cutter 351
        Procedure 7—Cutting Square Stock with a Hydraulic Cutter 352
        Procedure 8—Cutting Wood with a Reciprocating Saw 353

8—Entrapment beneath and between Vehicles 355
    General Procedures 357
        Hand tools and wedges 360
        Pry bars 360
        Hi-Lift jack 360
        Come-along 361
        Manually operated spreaders 361
        High-pressure air bags 362
        Heavy hydraulic spreaders 363
    Procedures 365
        Procedure 1—Lifting a Vehicle with Wedges 366
        Procedure 2—Lifting with a Pry Bar and Cribbing 367
        Procedure 3—Lifting a Vehicle with Manual Hydraulic Tools 368
        Procedure 4—Lifting with High-pressure Air Bags 369
        Procedure 5—Lifting with High-pressure Air Bags 370
        Procedure 6—Lifting with Hydraulic Spreaders 371
        Procedure 7—Lifting with Hydraulic Spreaders 372
        Procedure 8—Lifting on a Hard Surface with a Hi-Lift Jack 373
        Procedure 9—Lifting on a Soft Surface with a Hi-Lift Jack 375

Contents

    Procedure 10—Lifting with a Come-along in Soft Soil 377
    Procedure 11—Lifting an Overturned Vehicle with a Hi-Lift Jack 380
    Procedure 12—Spreading Two Vehicles with Hydraulic Tools 382

9—Trucks and Tractor Trailers 385
  Common Terms 385
  Popping a Door with Hydraulic Spreaders 388
  Performing a Dash Roll-up 390
  Trailer Underride by an Automobile 393
  Summary 393
  Procedures 395
    Procedure 1—Opening and Removing Conventional Hoods 396
    Procedure 2—Truck Stabilization 398
    Procedure 3—Removing a Windshield with Hand Tools 401
    Procedure 4—Popping a Door Open with an Air Chisel .403
    Procedure 5—Popping a Door Open with Hydraulic Spreaders 405
    Procedure 6—Removing a Door at the Hinges with an Air Chisel 409
    Procedure 7—Removing a Door with Hydraulic Tools 410
    Procedure 8—Pedal Removal with Pneumatic Tools 414
    Procedure 9—Pedal Removal with Hydraulic Tools 415
    Procedure 10—Cutting Away a Steering Ring with Hydraulic Cutters 416
    Procedure 11—Cab-over Dash Roll with Pneumatic and Hand Tools Only 417
    Procedure 12—Tenting a Roof with Hydraulic Spreaders 420
    Procedure 13—Roof Entry with a Reciprocating Saw 421
    Procedure 14—Roof Entry with an Air Chisel 424
    Procedure 15—Dash Roll-up with a Reciprocating Saw and Come-along 429
    Procedure 16—Dash Roll-up with Hydraulic Tools on a Cab-forward 435
    Procedure 17—Dash Roll-up with Hydraulic Tools on a Cab-over 437
    Procedure 18—Dash Roll-up with Hydraulic Tools on a Conventional 439
    Procedure 19—Dash Lift with Hydraulic Tools on a Cab-over 446
    Procedure 20—Cab-over Dash Removal with a Reciprocating Saw 449
    Procedure 21—Sleeper Side Entry with a Reciprocating Saw 452
    Procedure 22—Sleeper Side Entry with an Air Chisel 454
    Procedure 23—Rear Entry into a Conventional Steel Cab with a Reciprocating Saw . . .457
    Procedure 24—Rear Sleeper Entry with a Reciprocating Saw 459
    Procedure 25—Sleeper Rear Entry with Hydraulic Spreaders 461
    Procedure 26—Sleeper Rear Entry with Hydraulic Cutters 464
    Procedure 27—Trailer Underride 467

10—Extrication Procedures for School and Transit Buses 477
Procedures 483
Procedure 1—School Bus Emergency Exit—Side Door Exit 484
Procedure 2—School Bus Emergency Exit—Roof Hatch 485
Procedure 3—Transit Bus Emergency Exit—Roof Hatch 486
Procedure 4—Transit Bus Emergency Exit—Window 487
Procedure 5—Removing Windshield Glass with Hand Tools 488
Procedure 6—Removing Side Window Glass with Hand Tools 489
Procedure 7—Front Door Entry with Hand Tools 490
Procedure 8—Front Door Entry with a Reciprocating Saw 492
Procedure 9—Front Door Entry with a Hydraulic Cutter 494
Procedure 10—Emergency Exit Door Removal with an Air Chisel 498
Procedure 11—Popping an Emergency Door Open with Hydraulic Spreaders 501
Procedure 12—Removing the Rear Window Posts with a Reciprocating Saw 502
Procedure 13—Removing the Rear Window Posts with a Hydraulic Cutter 504
Procedure 14—Rear Wall Removal with Hydraulic Tools 505
Procedure 15—Roof Entry with a Reciprocating Saw 508
Procedure 16—Roof Entry with an Air Chisel 510
Procedure 17—Side Wall Removal with a Reciprocating Saw 512
Procedure 18—Side Entry with an Air Chisel 517
Procedure 19—Removing a Steering Ring with Hydraulic Cutters 520
Procedure 20—Relocating a Steering Column with a Come-along 521
Procedure 21—Driver's Seat Back Removal with Hydraulic Tools 525
Procedure 22—Seat Post Removal with a Reciprocating Saw 528
Procedure 23—Seat Removal with an Air Chisel 529
Procedure 24—Seat Removal with Hydraulic Tools 532
Procedure 25—Lifting a Bus with Air Bags 534
Procedure 26—Lifting a Bus with Hydraulic Spreaders 537

11—Sport and Race Vehicles 539
Construction Basics 540
Roll bars and cages 540
Fuel and fuel tanks 540
Steering wheels 541
Seats and belts 541
Doors and other means of access to the interior 541
Shutting down the engine 542
Summary 542
Appendix: Things to Think About 545

Index 557

Photo Credits

All photos were taken by the author, with the exception of the following:
- Fig. 1–2 Miami-Dade Fire Rescue Public Information Office
- Fig. 2–16 Photo #2 Miami-Dade Fire Rescue
- Fig. 2–19 Rick Michalo, American Rescue Technology
- Fig. 2–20 Rick Michalo, American Rescue Technology
- Fig. 2–30 Photo #4 Rick Michalo, American Rescue Technology
- Fig. 3–20 Frank Gentilquore, Zuccala's Wrecker Service
- Fig. 3–21 Frank Gentilquore, Zuccala's Wrecker Service
- Fig. 3–27 Miami-Dade Fire Rescue Public Information Office
- Fig. 3–28 Manny Santiestaban, Miami-Dade Fire Rescue
- Fig. 5–3 William Lehman Injury Research Center
- Fig. 7–1 Tyrone Wilson, Miami-Dade Fire Rescue
- Fig. 7–2 Tyrone Wilson, Miami-Dade Fire Rescue
- Fig. 7–3 Brian Gordon, Riviera Beach Fire Department
- Fig. 10–4 Miami-Dade Fire Rescue Public Information Office
- Fig. 10–5 Miami-Dade Fire Rescue Public Information Office

Acknowledgments

The author would like to acknowledge the following extrication professionals for their assistance with staging procedures throughout the text.

A. Alvarez

Shane Anderson

Mike Belle

Ken Brack

Phil Burden

Ben Campbell

Bill Conderman

Chris Couden

J. R. Diaz

Keith Earle

Jack Garcia

Mario Gonzalez

Brian Gordon

Brad Havrilla

Michael Humphries

Kenny Jones

Matt Mandel

Enrique Marino

John McLaughlin

Daryl Newport

Vehicle Extrication: A Practical Guide

Bernhard Obermayr

Cecilia Perales

Ray Ramirez

Gary Richard

Nick Rousseau

Hugh Saenz

Terry Salvi

Jose Santiago

Manny Santiestaban

Todd Schindler

Bill Serey

Michael Shively

Steve Shupert

Mike Slattery

Charles Sosa

Stephen Springs

Jeff Strickland

Chas Sunser

David Sweet

Jack Swerdloff

Acknowledgments

Craig Tamkins

Gio Ulloa

J. D. Vasbinder

Omar Vega

Brandon Webb

Kerry Weiss

Tom Wilcox

Tyrone Wilson

Henry Wong

Chris Zargo

Drew Zuccala

Acknowledgments

As I attempt to create an all-inclusive list of everyone who has helped with this project, I find it difficult to clearly identify when the groundwork that culminated in production of this book was first laid. Some of the best formal training I received on the subject occurred at the Florida State Fire College, where I had the good fortune to attend an extrication class taught by Bill Nesmith and Dave Sherratt. The class was based on a course developed by the Ontario Fire Marshall's Office, and it provided me with a solid foundation based on proper tool usage and practical procedures. Since that time, a continuous exchange of procedures and philosophy with students, instructors, and other members of the fire service has resulted in a deeper knowledge in the field of vehicle extrication and the production of this book.

Like any large fire department operation, this book was made possible only by the cooperative effort of many individuals and organizations, all willing to do their parts to bring the project to a successful conclusion. As the last words are typed and the last photos are taken, I find myself reflecting on the amazing and enormous contributions of others, for without their help, this book would have remained a concept instead of a reality.

The greatest contributions undoubtedly were made by those who helped set up and demonstrate each step of each procedure displayed in the following pages. Their good nature and willingness to stand perfectly still for thousands of photographs made that part of my job easy.

For chapter 9, "Trucks and Tractor Trailers," I received a tremendous amount of assistance from J. D. Vasbinder and Tom Wilcox, two gentlemen whose depth of knowledge on the subject is only surpassed by the depth of their humility.

Other members of the fire service who helped in a variety of ways include Gary (HB) Klaus, Bill Gustin, Louis Fernandez, Chris Seiler, Doug Lavalley, and Wes Roberts.

The vehicles and props used in the photographs were provided by the same individuals who have always been there for us when it has been time to perform extrication training. This group of fine individuals includes Paul Smith, Bob Kaplin, Marvin Ward, and Bob Reece of Reliable Truck Parts, Columbus, Ohio; Bruce White of Sadisco, West Palm Beach, Florida; Phil Everett, John Everett, and John Mick of Recycling Center, West Palm Beach, Florida; Greg Ledet, Jeff Horowitz, and Ruffo Escorcia of Trademark Metals Recycling, Opa Locka and West Palm Beach, Florida.

In West Palm Beach, where much of the photography was done, Val Williams, Bob Brown, and Tim Monaghan of the Palm Beach Community College Fire Academy provided assistance with facilities and tools that was very beneficial and greatly appreciated. While working on chapter 9 in Ohio, Frank Conway of the Ohio Fire Academy provided logistical support with their state-of-the-art training vehicle. I am thankful for the support from the leadership of both of these fire academies.

Vehicle Extrication: A Practical Guide

Help with the special problems associated with police vehicles and the application of new hybrid vehicle technology was provided by Sheriff Robert Crowder, Jenell Atlas, and Carl Amerson of the Martin County (FL) Sheriff's Office, and Mike Stewart of Pro-Gard Police Products, Indianapolis, Indiana.

Specialized information about vehicle design and operation was provided by Eric Bowman of Sage Technical Services; Howard Veit of Heavy Truck Collision Center in Riviera Beach, Florida; Rick Capri of Community Transit in West Palm Beach, Florida; John Hauser and Keith Sorrell of Elite Paint and Body Shop, Riviera Beach, Florida; and Kris Foraker of Wayne Akers Ford, Lake Worth, Florida.

Tools were generously provided by Don Harrison of Rescue Systems Inc., Panama City, Florida; Rick Michalo of American Rescue Technology, Dayton, Ohio; Todd Howell of Howell Rescue, Dayton, Ohio; James Riddle of High Tech Rescue, Shelbyville, Kentucky; Bernhard Obermayr of Weber-Hydraulic GMBH, Losenstein, Austria; Boyd Miller and Gary Parker of Milwaukee Electric Tools, Brookfield, Wisconsin; Kim Silcox of Grainger Industrial Supply, West Palm Beach, Florida; Steve Dowden of Hi-Lift Jack Company, Bloomfield, Indiana; Dave Taylor of Taylor Pneumatics, Boynton Beach, Florida; Rick Kozub of Lug-all Corporation, Morgantown, Pennsylvania; and Baldor Electric Company, Fort Smith, Arkansas.

Ideas for storage and deployment of equipment came courtesy of Fire Chief Ray Carter and Chris Couden of West Palm Beach (FL) Fire Rescue and Fire Chief Bobby Cowherd of Shelby County (KY) Suburban Fire District.

The capability of large wreckers was explained by Drew Zuccala and Frank Gentilquore of Zuccala's Wrecker Service, Boynton Beach, Florida.

Information regarding race cars was provided by Clayton Murphy, George Magale, and Danny Fournier of Chassis Engineering, Riviera Beach, Florida; Jimmy Adams of Miami-Dade Fire Rescue; Robert Chandler of Bigfoot 4X4 Inc., St. Louis, Missouri; and Paul Brennan and his Xtreme Funny Car, based in Palm Beach County, Florida.

This book took more than twice as long to write as I originally thought, making me appreciative of the support of Jared Wicklund, Supervising Editor at PennWell, who patiently kept the whole book project under control and moving forward.

Above all, I am thankful for the support provided by my wife Georgia while I worked on this book. She's been there from the beginning, providing advice about the organization and layout of the book and then proofreading every word of text. Her greatest contribution may have come at the very beginning when she said, "If you're going to write a book about extrication, why don't you do it for *Fire Engineering*?" What a great idea.

Acronyms

ALS – advanced life support

BLS – basic life support

BPM – blows per minute

cfm – cubic feet per minute

CGA – Compressed Gas Association

EMT – emergency medical technician

ETA – estimated time of arrival

HAZMAT – hazardous materials

kPa – kilopascals

LZ – landing zone

ma – mechanical assistance

MCI – mass casualty incident

NFPA – National Fire Protection Association

PPE – personal protective equipment

PTO – power take off

psi – pounds per square inch

SCBA – self-contained breathing apparatus

SI – international system

SIPS – side-impact protection system

SRS – supplemental restraint system

TPI – teeth per inch

Getting Organized

SHARP FIREFIGHTERS, POWERFUL TOOLS, AND STRONG LEADERSHIP

Engine 30, respond with Engine 32 to I–95 and Northwest 135th Street for a report of traffic accident with persons trapped…standby…Engine 30, FHP now confirms persons trapped and the vehicle is on fire…FHP requesting an ETA.

This type of dispatch occurs on a regular basis across the United States in small towns, in the suburbs, and in big cities. Extricating trapped victims is the most common type of rescue performed by firefighters. It is also a time when the skill of the firefighters has a significant impact on the survival of the victims.

Many of the extrications performed by firefighters are simple, intuitive operations, requiring nothing more than popping the door to release the occupant. Others, though, can be more challenging, and occasionally a crash will test a firefighter's limits. These are the crashes that are so bad and the entrapment so complicated that even experienced firefighters have to stop and take a few seconds to try to understand what they're seeing. These crashes evoke comments like, "Does this car have two doors or four?" "How many victims can you see from your side of the car?" "Where's the front half of the car?" and "This is really bad." These serious crashes require the highest level of skill from the firefighters, great power from the tools, and strong, confident leadership from the officers.

When the company possesses these traits, along with solid teamwork and just a little good luck, you've got a crew that can make even the most difficult extrications look easy.

Bad crashes are challenging because several procedures are usually needed to free the victims. Firefighters must use combinations and variations of procedures to remove the odd configurations of metal that wrap around the unfortunate occupants. However, before a crew can modify or combine procedures, they must have a broad knowledge base and skill level of the standard procedures. This is when training comes into play. Just like hose and ladder work, practice makes perfect—or at least pretty close. Crews can attain maximum effectiveness at the crash scene through preparation: studying procedures, training with tools, and practicing crash scene scenarios.

However, things don't always work as they do on the drill ground—even when the crew is well trained. In the real world, hose lines burst, power plants don't start, and sometimes tools just lack the power they had when they were new—or they fail. For reasons like these, it's important to have multiple solutions available to the crew. Understanding that there may be multiple solutions for a single problem can help a crew overcome any little snag that occurs during an operation. It is important that crews develop the knowledge and skill needed to adapt to the challenges and surprises of crash scenes.

Gaining Knowledge and Skill

Some procedures are so simple that common sense alone can guide even the most inexperienced firefighter to a successful conclusion. However, as the procedures become more complex, intuition may fail us due to the number of variables. *How can the metal be cut away from the victim's feet when the feet can't be seen? Why are the vehicle safety structures that are supposed to help protect the occupant hurting the occupant when the door is forced open? When metal starts to tear, is it a bad thing—or a good thing?*

There are basically two approaches for dealing with the learning curve of vehicle extrication: reinventing crash scene procedures, or learning from experienced firefighters. Most everyone would agree that learning from those who came before us is preferable. This provides more time to create new procedures and to fine-tune old ones.

This book attempts to illustrate multiple solutions to each type of problem that may be encountered by crews in the street. There is a solution for just about everybody—regardless of the type of tools they carry on their rigs: hand tools, pneumatics, hydraulics, or electric tools. Once a company learns the procedures and is ready to respond, they should focus on organizing themselves in a way that they can operate effectively while applying their skills. This involves understanding everyone's role, why they have those roles, and how they contribute to the team effort needed for top-notch operations.

Role and Responsibility of the Company Officer

Understanding, assigning, and carrying out the roles and responsibilities are critical steps for effectively managing a crash scene—especially those involving the extrication of victims. The early decisions and actions of the company officer and firefighters have a direct impact on the effectiveness of the rescue operation and victim survival. Strong leadership, coupled with sound decision-making and teamwork can result in a scene that proceeds in a coordinated, calm manner. Job responsibilities are assigned based on the structure of the organization, personnel, equipment, and experience. While the number of personnel and their titles vary, the needs, decision-making process, and actions remain the same. At small incidents, the company officer may be directing a crew and managing the overall scene. At large, complicated, or exceptionally long extrications, the company officer should transfer command of the incident to a higher ranking officer, allowing the company officer to concentrate on the extrication without the additional responsibilities of managing the entire scene.

Making tough decisions

In life, for every action there is an equal and opposite reaction. This is not just a difficult physics theory but a simple fact of life that can be observed on a daily basis. Stay in bed all day, and you may avoid being hit by a bus. By the same token, staying in bed all day may lead to bedsores, heart disease, and early death. Life is full of decisions, and life on the crash scene is no different.

Every action on the crash scene impacts someone or something connected to the crash. The multiple factors at a crash scene (including inadequate information) can mentally overload personnel, causing decision-making to be difficult. This is the nature of the fire service. In these situations, officers and firefighters alike have to make decisions based on available information, experience, and training. Decision-making under these circumstances can be stressful, especially for individuals who are newly promoted or lack experience, and this is normal. Nonetheless, at crash scenes, decisions must be made—no matter how difficult. In the words of Captain Joe Price (ret.), "If this was easy, everybody would be doing it."

Size-up

The company officer is responsible for developing a plan to bring incidents to a successful conclusion. As with all incidents, the plan starts with the initial dispatch information. This is the officer's first chance to make decisions that impact the management of the crash scene. When dispatched to an incident, the smart company officer does more than answer the radio, don bunker gear, and climb on the apparatus. The officer should also listen carefully to the dispatch information and determine if the appropriate resources have been dispatched to the incident. Initial dispatch assignments are simply that—initial. After the dispatch, the company officer may find it necessary to make changes in the initial assignment. Company officers who rely solely on dispatchers to make all the decisions about all assignments will eventually come up short on a response. Dispatchers should not be held responsible for all of the company officer's anticipated needs, especially when the dispatch center is busy. In most cases, it is best to consider the dispatcher's actions as part of the overall team effort required for successful operations.

The following are examples of situations that may require upgrading or modifying an initial dispatch assignment:

- Limited access highway (freeways, interstates, etc.) crashes
    - Additional unit to help block scene the from on-coming traffic
    - Faster response by other units based on access or direction of travel
    - Additional units for tank water when a vehicle fire is reported
- Victim information received while en route to the scene
    - Confirmed multiple victims
    - Confirmed critically injured victims
    - Confirmed entrapment
    - Ejected victims
- Terrain and geography
    - Submerged vehicles
    - Suspended off bridges, overpasses
    - Off-road in mountainous areas
    - Distant locations, which result in extended response times

Size-up upon arrival

After arriving at the crash scene and working with other company members, the officer should perform an initial size-up to determine if adequate resources have been assigned to the incident. The adequacy of the resources is based on the following:

- Scene hazards
- Number of victims
- Severity of the victims' injuries
- Number of entrapments
- Complexity of entrapment

The process of gathering information about the scene and victims is handled best if the officer assigns portions of the scene to the crew members on the unit. One of the most widely accepted methods of scene size-up involves dividing the scene into two areas: the inner circle and the outer circle. By dividing the scene into two areas, the company officer can make size-up assignments to as few as two crew members and still cover the area in a brief amount of time.

The inner circle can be difficult to describe, and the description may be vague because of the differences in crash scenes. The inner circle can generally be thought of as the area under and around the vehicle, within a radius of about 10 ft, or that portion of the scene that would come into the field of vision of a crew member who is within 5 ft of the vehicle. The inner circle can also be described as the area that can only be evaluated by the crew member who is standing next to the vehicle.

The outer circle is considered as all that *isn't* the inner circle. The outer circle extends from the edge of the inner circle to the outer edge of the

Getting Organized

crash scene. Determining the outer edge of the crash scene can be difficult at times, especially when the vehicles were traveling at a high rate of speed, when the crash occurred in poor weather, or at night when visibility is limited. When establishing the boundary of the outer circle, the crew is making a judgment call that all vehicles, victims, and hazards will be found within the inner and outer circles.

Presence of hazards

Hazard recognition can be difficult, especially at night, during bad weather, and at large incidents. The company officer must focus on getting the big picture—a solid understanding of what is going on at the scene. As the officer pulls up to the scene, the next phase of hazard recognition begins, picking up where the original dispatch information ended. The officer should address the hazards created by the following:

- Traffic.
- Damaged utility service.
- Vehicles involved in the crash.

A plan should be developed and implemented to mitigate or stabilize the existing hazards. The process of hazard mitigation and stabilization is covered in chapter 3.

Determining adequacy of resources

Medical units. One of the difficult decisions the first arriving officer must make on the scene is the number of medical units required for treatment and transportation of victims. Following are factors that impact this decision.

- Are the responding medical units advanced life support (ALS) or basic life support (BLS)?
- How many personnel are on the medical units?
- What is the medic crew's degree of involvement in performing the extrication?
- Is the area served by a trauma center and/or trauma helicopter?
- What is the severity of injuries?
- Most importantly, how many victims are there?

As the number of victims and severity of injuries increase, so does the likelihood of discovering too late that inadequate resources have been assigned to the incident. Depending on the type and number of resources available, the officer should decide whether the incident should be handled as a mass casualty incident (MCI) or as an ordinary crash incident.

MCI is the term used to identify incidents when the sheer number of victims overwhelms the initial units, requiring the assignment of additional units and the application of procedures that promote the rapid assessment, treatment, and transportation of the injured. If only a few victims are injured, and the incident is to be handled in a routine manner, the officer must determine the number of units required to treat and transport the victims.

One easy way to handle the math in this type of situation is to consider how many personnel and units would normally be required to adequately and properly handle one critically injured crash victim. For example, if it would normally take one medical unit per critical victim, and there are three critical victims on the scene, three medical units are needed. If this procedure for determining the number of units needed seems high, consider that with inadequate medical resources on the scene, somebody is not going to be treated in a timely fashion. If the situation requires the dispatch of mutual aid, so be it. On scenes with multiple victims, we can hope to get the victims off the scene quickly, but the reality is that, without adequate resources, treatment gets bogged down along with transportation to the receiving hospital. This applies to both MCIs and ordinary crash scenes.

In many cases, crew members assigned to a medic unit are committed to the assessment and treatment of the trapped victim from both inside and outside the vehicle. These crew members should wear bunker gear, which provides good personal protection. If additional resources are available, a replacement medic unit should be used for transportation from the scene, relieving the original crew (whose bunker gear is cumbersome to work and drive in) from that responsibility.

Hospital or trauma center assistance. Occasionally even the best-trained and most experienced extrication crews are confronted with an entrapment for which there is no good, timely solution. These extrication scenes may drag on for hours, grinding down the personnel charged with the task of disentangling the victims. In some cases, the occupant may not have any significant injury and may remain medically stable throughout the incident. On the other hand, there are situations when the crash victim is critically injured and may require advanced treatment from hospital or trauma center personnel during the extrication process. Victims with the following needs may benefit from the involvement of hospital or trauma center personnel:

- On-scene administration of blood
- Amputation due to entrapment
- Completion of near-amputations
- Procedures to relieve tension pneumothorax (collapsed lung)

To successfully utilize hospital or trauma center personnel on a crash scene, policies, procedures, and agreements should be formalized before an incident occurs that requires these types of resources. Once the policies are in place, all personnel who are affected by the policy should know their role and responsibilities at a crash scene (see Fig. 1–1).

Suppression units. The role of suppression units at crash scenes varies widely across the country. In some areas, engine companies are responsible for extrication; in other areas, truck companies handle the task. It is becoming common for suppression units to be staffed with personnel that are not only firefighters, but emergency medical technicians (EMTs) or paramedics as well. Suppression units, whatever their form, generally have several responsibilities on a crash scene, which include the following:

- Scene safety and management
- Fire suppression
- Extrication of trapped victims
- Assistance in medical treatment
- Helicopter landing zone (LZ) set-up

Getting Organized

MIAMI DADE FIRE RESCUE
MEDICAL OPERATIONS MANUAL

FIELD SURGICAL KIT / AMPUTATION

PROTOCOL 31

A. Introduction
In the rare event of a patient entrapment in which all resources have failed to successfully extricate the patient, or in which the patient's life is in immediate danger with prolonged extrication, field amputation of the trapped extremity should be considered. The Field Surgical Kit also carries equipment necessary to perform surgical airways, insertion of a chest tube, and placement of central venous access.

B. Procedure for Requesting

General Care

1. Prior to instituting a request, ensure that a TRT unit has been dispatched and is en route.
2. Either the Rescue OIC or Incident Commander can request that the Fire Alarm Office notify the Medical Director or Deputy Medical Director of the need for a possible field amputation or other advanced procedure.
3. The Fire Alarm Office will make contact with a Medical Director as quickly as possible. Only the Medical Director, Deputy Medical Director, or their appointed representative, is authorized to conduct this procedure.

    NOTE: In the event the Medical Director or the Deputy Medical Director is not available for transport to the scene, either may request the assistance of the Ryder Trauma Center Attending Trauma Surgeon or their designee.

4. The identified physician will be transported to the scene in the most expeditious manner after consultation with the Incident Commander.
5. The Incident Commander or designated physician must confirm that at least one Field Surgical Kit is available at the scene. Sources for a Field Surgical Kit include:

    a) One kit is located at each Air Rescue station
    b) One kit is located at the JMH/Ryder Trauma Center
    c) One kit is carried by the Technical Rescue Bureau OIC

6. Once on scene, the physician will assess the situation at hand and make the decision regarding the requirement for amputation. The physician on scene will have the final say as to the appropriateness of a field amputation.

02/03 PROTOCOL 31.1

Fig. 1-1

Specialty units. During the crash scene size-up, the company officer should determine if the units assigned to the incident can handle the crash and the extrication. This decision should be based on the ability of the personnel, the complexity of the extrication, and on the equipment needed to perform the required procedures. As soon as it becomes apparent that the on-scene resources can't perform the tasks required, specialty units should be requested. Specialty units may be in the form of squads, technical rescue units, or heavy rescues—based on regional needs and tradition. These units should be equipped with the tools and personnel needed to perform specialized or complicated extrications.

Wreckers. When wreckers are needed to assist with rescue operations, they should be requested early in the incident. Routine towing requests are typically handled by law enforcement personnel and may be handled differently than requests for assistance to perform a rescue. When requesting a wrecker, the officer should clearly indicate that the request is needed for a rescue, and that it is not a routine towing request. By making the request clear to the dispatcher, there is less chance of the request being handled as a routine towing request (see Fig. 1–2).

Developing the tactical plan and assigning tasks

When developing a tactical plan and assigning tasks, a company officer can be overwhelmed by the multiple problems that need to be resolved before the situation can be brought to a successful conclusion. Tactical priorities must be assigned, but the question is *where to begin*? While every scene is different, the following approaches may be helpful when selecting a course of action:

1. Actions that stop the incident from getting worse—if a vehicle has struck a tree and a small fire has started in the engine compartment, the fire usually needs to be extinguished before any victim treatment begins. The fire is a factor that can cause the incident to deteriorate if left unchecked.

2. Actions that will do the most good for the most victims—if all the victims are in a similar medical condition, actions that free the greatest number of victims are most appropriate.

3. Actions to extricate the most severely injured first—if given a choice between extricating an unconscious victim and a conscious victim with a cut arm, extricate the unconscious victim first.

These guides are helpful for basic decision-making, but local capabilities and resources ultimately dictate the type of approach, especially at MCIs or disasters.

When there are overlapping needs, Chief Daryl Newport of Palm Beach County (FL) Fire Rescue, suggests that usually 90% of the problems on a crash scene can be resolved by one or two actions. Consider the example of a car that has rolled over several times and has come to rest on its side with five teenagers trapped in the vehicle. The roof side of the vehicle is exposed to the crew, and the frame side of the vehicle is against a tree. As the victims move around inside the vehicle, the vehicle rocks, nearly rolling over onto the flattened roof. Assessment is difficult because the flattened roof makes it almost impossible to see

Getting Organized

This vehicle is classified as a 50 ton, Class D Recovery Vehicle.

When not in use, the under-reach component is stored in the upright position.

The under-reach can be used to lift heavy vehicles like buses and trucks.

The under-reach can be extended and lowered to ground level when needed.

Frame forks can be attached for a more secure lift.

Chains supplement the forks.

Fig. 1–2 Overview of a Heavy Duty Recovery Vehicle

the occupants, let alone treat them. It becomes apparent the roof must be removed to release the victims, but cutting the posts will certainly cause the vehicle to roll over onto the roof. *Which one or two actions can be taken to make the situation better?*

If the officer orders the vehicle to be tied back to the tree, potential problems are prevented. The odds of the vehicle rolling over are greatly reduced once the vehicle is tied to the tree with a rope, strap, or chain. With the addition of a

Vehicle Extrication: A Practical Guide

Side view of the under-reach fully extended.

The boom pistons are used to lift.

To reach out for an overhead lift, the boom can be extended.

The boom can also be elevated in the extended position.

The end of the boom is outfitted with dual boom sheave cable guides.

The cable guides pivot, allowing pulling procedures to be performed off the side.

Fig. 1–2 *(cont.)* Overview of a Heavy Duty Recovery Vehicle

little cribbing between the ground and the body of the vehicle, there's a good chance the roof can be removed safely, providing access for victim treatment and removal. In this case, stabilizing the vehicle and removing the roof solve the vast majority of the problems on the scene.

Getting Organized

Hooks with latches are attached at the end of each cable.

For 2:1 mechanical advantage, a block can be attached to the cable.

In this example, the cable guide swivels up, allowing the cable to pay off the rear.

When using the block, rear mounted recovery anchors are used.

The cables are powered by dual winches.

The operator can use truck mounted controls or remote contols to control each function.

Fig. 1–2 *(cont.)* Overview of a Heavy Duty Recovery Vehicle

The basic tactical options

Meaningful tactical solutions and decisions can be made only after first identifying the problem. Before the officer and crew can make decisions about the tools and procedures needed to extricate a victim, they should first answer the simplest, yet most critical question: *What is trapping the person inside the car?* The officer and crew must aggressively examine the deformity of the vehicle and then identify the components that are preventing the victims from getting out of the vehicle. As the severity of the crash increases, the ability to even identify the components of the vehicle decreases, making it difficult for the officer to develop a plan. If a vehicle is involved in a severe crash, it may be difficult to determine where the dashboard ends, where the door begins, or how many doors are on the vehicle. After quick and careful examination of the vehicle and the trapped victim, the officer and crew can select a course of action to free the victim.

When selecting the best tactical option for an extrication, how does the officer know if all the options have been considered? In other words, how do you know if you didn't think of something? This uncertainty can be reduced by remembering that almost all extrication procedures are based on one of following four tactical options.

1. Spreading—the process of spreading or pushing apart is the most common action used on the crash scene. The adz (flat portion) of a Halligan is used to create a purchase point for the spreaders by spreading the door metal away from the body of the vehicle until the door is forced open. Rams are used to spread or push the dashboard out of the occupant compartment.

2. Pulling—components are pulled apart when the space between two objects is too large or too flimsy to be spread or pushed apart. A come-along can be used to pull a door open when the space created by the spreaders is inadequate to gain access to the occupants. A front seat may be pulled rearward to gain access to an occupant who is trapped on the floor under the dashboard.

3. Removing—when pushing or pulling an object won't provide the space needed to access or remove victims, severing and removing the object may be the only other choice. If a victim is trapped in the rear seat of a two-door vehicle, spreading or pulling isn't the best approach to creating a big space. In this situation, performing a third-door evolution with an air chisel is the preferred procedure.

4. Securing—preventing vehicle movement may be required in some situations to prevent fatal injuries. For example, if a passenger has been partially ejected from the car, and the vehicle has pinned the occupant between the ground and the vehicle, any additional movement could cause fatal injuries. In this case, it's necessary to secure the vehicle and prevent even the slightest movement.

To bring this concept of four tactical options into focus, consider the procedure of dealing with a badly damaged door. To gain access to the occupant in the front seat, the officer has a choice of three of the four standard tactical options:

1. Spreading—after sizing up the vehicle, the officer can direct the crew to use the spreaders to *spread* the door off the latch. With the door popped off the latch, the operation can be continued to push the door open, overcoming any resistance caused by the damage to the hinges. The spreading option can be used until the spreaders are opened completely.

2. Pulling—if the spreaders haven't created enough space, the *pulling* option can be used by attaching the come-along to the door and an anchor point.

Getting Organized

3. Removing—if the latch or hinges are exposed, the shears may be used to cut through them, and the door can be *removed* from the vehicle.

Before committing to and implementing a particular procedure in complex extrications, the officer should quickly visualize the procedure step by step to determine if it leads to the right outcome. By performing the steps of the procedure mentally, the officer may be able to identify problems before they occur or may see that the planned procedure simply won't work. This mental process must be done quickly if it is to be valuable. If the plan seems likely to work, the officer can brief the crew on the tactics to be used to perform the extrication. Briefing the crew will provide them with a critical understanding of the objectives and an opportunity for a crew member to identify any possible problems.

To understand the value of running through the procedure mentally before committing to it physically, consider the following scenario.

A compact car leaves the roadway and rolls over several times coming to rest on its roof. As the vehicle rolls, the unbelted driver is partially ejected through the windshield. The roof is pushed down and the driver is lightly trapped between the roof and the dashboard. The right front seat passenger is belted and remains unconscious in his seat, with his head pressed against the roof. After sizing up the entrapment, the officer decides to quickly cut the roof off the car to take the pressure off of both occupants. Only when the officer runs through the plan mentally does it become apparent that cutting the roof posts will remove the very components that are holding the vehicle off of the victims. If the roof posts are cut and can't support the car, the car would fall flat creating a bad situation for the occupants.

When performing simple procedures such as popping a door, the mental process of running through the procedure can be completed in a matter of seconds. Experience and training allow the officer to predict the limited number of outcomes associated with the procedure, making the process almost unnecessary. The process becomes much more important, however, when the extrication is complicated or when a vehicle comes to rest on its side or roof. These are the situations that can result in tactical errors, validating the use of the mental process before applying the physical process.

Risk/benefit analysis

There are times when the officer has a choice of several procedures to use in a particular situation. There may be drawbacks to all of the choices, each possessing a certain element of risk. The question then becomes, *how does one choose the right procedure?*

Risk/benefit analysis is an important process when making decisions at emergency scenes. Risk/benefit analysis means weighing the risks associated with performing a procedure against the benefits that can be gained from the procedure. In simple operations, risk/benefit analysis is often performed without much thought because the risks and benefits have already been analyzed during training or other similar situations. As situations become complex or unfamiliar, it is acceptable and appropriate to quickly and methodically analyze the options available along with the associated risk and potential benefits.

Officers with the ability to make an accurate risk/benefit analysis perform much more effectively than those who believe that there is only one *approved* procedure for every situation. Over the past 25 years, the quality of both tools and procedures has improved dramatically. Along with these improvements has come the idea that in every situation there are necessary procedures that must be performed, or the extrication crew could be considered negligent. This, of course, is not true since circumstances dictate actions, and the officer must choose the best, real-world solutions for the problem at hand.

To clarify the relationship between risk and benefit, consider the following question: *Is it appropriate to use a ventilation-type saw fitted with a metal cutting blade to perform an extrication?* In areas of the country that have hydraulic rescue tools, most would say that it is never appropriate to use the saw. Saws can be difficult to control for the inexperienced operator; they produce sparks that can start fires, and they throw metal slivers. Considering the significant downside of using the saw when hydraulic tools are available, it seems that the risk of using the saw far outweighs the benefit. The balance of risk/benefit may change however, if an officer is in charge of an extrication with a person trapped and the nearest hydraulic tool is 80 miles away. If the success of the entire rescue and survival of the victim is based on cutting one hinge to free the victim, the officer may decide that the benefit of using the saw far outweighs the risk. If the decision is made to use the saw, the officer is accepting the risk inherent with the operation.

When making these kinds of decisions, a risk/benefit analysis can provide the officer with the mental and psychological advantage needed when faced with a particularly difficult challenge. When a risk/benefit analysis is combined with strong leadership skills, tactical paralysis can be avoided in even the most complex incidents.

Plan A, B, and C

Once the decision has been made about how to proceed and the crew has started working, the officer should develop a back-up plan in case the primary plan fails. The back-up plan is often one of the tactical options that was previously considered during the initial phase of the operation. To formulate Plan B, it's necessary to determine where and how problems in Plan A can occur. Problems can arise for several reasons.

- The tool fails to complete the operation. Examples include hydraulic tools that are worn out, poorly maintained, underpowered, or pneumatic tools that run out of air.

- The vehicle structure reacts in a manner that prevents completion of the procedure. Examples include components that are heavily reinforced (strong), or rusted away (weak), or metal that bends or tears in places that weren't anticipated.

- The procedure results in a negative impact on the victim trapped in the vehicle. Consider a severe side impact resulting in a deep V in a door. Attempts to pop the door open using the hydraulic spreaders may push the door in on the victim, causing additional injury.

Getting Organized

When developing Plan B, the officer has three options.

1. Modifying and improving Plan A
2. Selecting a different procedure using the same tool system
3. Selecting a different tool or combination of tool systems for the procedure

Consider a crash in which the officer has decided to remove the roof from a vehicle that has a large "C" post. The hydraulic cutters are used to cut the "A" and "B" posts, but when the firefighter attempts to cut through the large "C" post, the depth of the cutter only permits a cut 6 in. deep into a "C" post that is 18 in. wide. What should the officer do? If the department has the luxury of having different types of hydraulic cutters on the unit, switching to a cutter that has a long cutting blade may solve the problem. The firefighter would cut from one direction, then from the other direction to complete the cut. Without the luxury of having two types of cutters, the officer may decide to use a reciprocating saw instead of the hydraulic cutters. These are two solutions to one problem; one solution used the same hydraulic tool system, the other solution used a different electric tool system.

As the officer observes the progress of the extrication, he or she must decide to either stick with the original plan or go to Plan B. If the officer elects to change tactics, it's important that all personnel are aware of the change in plans. If someone doesn't get the word, unnecessary procedures may be performed that could waste critical time.

A tool staging area should be set up with all the appropriate tools needed for the change in tactics. At this point, the firefighter can go to the tool staging area and swap out tools as necessary with little loss of time. If Plan B is put in action, the officer should start considering Plan C. This may seem a little excessive, but it is necessary to have a back-up plan in place in case Plan B doesn't work out. If Plans A or B don't work and it's clear to the firefighters on the scene that the officer doesn't have a new plan prepared, a free-for-all could erupt on the scene, especially if the crew is getting fatigued and frustrated. At this time it's critical that the officer provides strong leadership and command presence to keep the crew on track and focused.

Extrication time estimates

After the firefighters have closely observed the victim entrapment, the officer should ask about the degree of entrapment and anticipated difficulty in relieving any entanglement. This information helps the officer when coordinating and planning for the operations of several units.

Accurate time estimates aren't usually a big factor when the victim is to be transported by ground units to the hospital or trauma center, but the accuracy of these estimates can make or break an air transport operation.

The following examples are useful in illustrating the importance of accurate time estimates when the victim is to be transported to the trauma center by air:

1. After one of two trapped victims is extricated from a vehicle, a ground transport unit transmits that they are en route to the LZ. Knowing that there are two victims that are to be transported to the trauma center, the officer on the trauma helicopter asks the officer at the scene for an

estimated time of arrival (ETA) for the second victim.

2. As the helicopter is lifting off the pad at the trauma center after clearing another call, they are requested at an extrication scene 30 miles away. The helicopter crew starts flying in the direction of the crash, but asks the company officer at the scene for an ETA to the LZ by the ground transport unit.

3. After sitting in the LZ with the engines running for 20 minutes, a helicopter crew member asks for an ETA to the LZ.

The important factor in all of these cases is the quantity of fuel on the helicopter. While ground fire apparatus can operate for hours without refueling, helicopters typically don't have that luxury. Accurate time estimates are important to the aircraft commander because there is a fixed amount of time that the aircraft can be safely operated, depending on the amount of fuel on board the aircraft. So the questions about the ETA to the LZ are not the product of curiosity but instead a factor in determining if the aircraft will have enough fuel to complete the mission. If the helicopter is being dispatched from one call to the next on a busy day, accurate time estimates can help the aircraft commander decide if it's best to stop and get fuel, or proceed directly to the LZ.

After arriving at the scene, the pilot may decide to conserve fuel by shutting down the engines on the aircraft or, if provided with a short ETA, may decide to keep them running. In all of these situations, the company officer and firefighters have a significant responsibility to the aircraft crew and victims when providing time estimates for completion of the extrication and estimated time of arrival at the LZ.

Things the company officer shouldn't be doing

When a firefighter is promoted to company officer, there is a distinct change in the individual's job description. The job description for company officer usually includes phrases like *evaluate situations, formulate plans, supervise,* and *direct personnel.* Typically there is no mention of *prying doors open, breaking glass,* or *striking with a sledgehammer.* There is a good reason for this: someone needs to be in charge at emergency scenes. Put in the simplest way, most of the time, the company officer should be thinking and directing—not handling tools. This isn't a new concept; it's been around a long time.

The traditional fire service insignias used to indicate rank support this idea. A firefighter's insignia usually includes a hook and a ladder representative of handling tools and equipment. The driver's insignia often includes a vintage fire engine. The company officer's insignia consists of a speaking trumpet. As the rank advances from lieutenant, to captain, to chief, the trumpets on the insignia increase in number. These insignias indicate that the company officer's role is to think and speak, not to handle tools. That's not to say that an officer can't ever handle tools, because that's not the case. There is a problem, however, when three firefighters are standing around watching an officer operate a set of hydraulic spreaders on a jammed door, or when a scene spins out of control for lack of leadership.

The implications of the officer performing a firefighter's role on the scene can have other negative effects beyond poor scene management. Firefighters may think the officer is "cherry picking," or choosing tasks and responsibilities based on personal satisfaction and not for the benefit of the victims or members of the company. In most cases, the company officer is most effective when concentrating on guiding crew members as they perform the tasks required at the crash scene.

Role and Responsibility of the Driver Operator

The driver of a suppression unit has three or four jobs at a crash scene.

- Spotting the apparatus at the scene to provide both good scene protection and easy deployment of extrication tools and fire suppression equipment
- Operating the pump when the company officer orders hose lines to be pulled and charged
- Setting up and staging equipment for the other crew members performing the extrication
- Operating the tools when directed by the company officer

The driver operator has a tremendous amount of work to do during the first 10 minutes on the scene. Training and experience help the driver anticipate the order in which tools will be needed, and this determines the sequence in which they should be set up. A tool staging area should be established in an area that is easy to identify, reasonably close to where the work is being performed, yet far enough away not to cause tripping hazards.

The concept of tool staging has been around for a while, but is often dismissed as "Hollywood" or excessive in layout. This is probably because depictions of tool staging areas in training materials seem to spend a little too much time and effort laying out the area. In reality, a tool staging area can be set up quickly by an organized driver, allowing other crew members to work faster. By utilizing a tool staging area, crew members won't have to struggle with getting equipment out of overstuffed compartments when they need a tool. Additionally, when a crew member is finished with a tool, there is a convenient, reasonably secure place to put it. This reduces the likelihood of a crew having to go to a towing company's holding yard to retrieve a tool that was laid in a vehicle during an extrication or turning around to grab a tool only to find that a bystander has walked away with it.

To help identify the location of the tool staging area, a tarp can be laid out and the tools placed on it. While there is no standard or correct size of tarp, a good size to start with is about 10 x 10 ft. The tarp should be a manageable size—easy to lay out, pick up, and able to hold all the tools that the crew normally uses at an extrication scene. This is a good use of old salvage covers that have been removed from the apparatus because of tears or burns. A good section of the old salvage cover can simply be cut to the desired dimensions of the tool staging area. Whether the tarp is made from an old salvage cover, or purchased specifically for tool staging, it should be made of heavyweight material so it will lay flat and resist being blown away. Ideally, the tarp is stored in the compartment with the other extrication tools to not only speed up the operation but also to remind the driver to establish the tool staging area.

Once the initial tasks have been completed, the driver can begin other non-critical but extremely helpful support functions, such as setting up scene lights, or spreading oil absorbent (if there are oil leaks) in the work area. By spreading absorbent, crew members working around the vehicle are less likely to slip and fall, and tools can be kept oil-free, making them easier and safer to use. During extended operations, the driver should monitor tool systems and maintain fuel and air supplies as needed.

Role and Responsibility of the Firefighter

If viewed superficially, many would think that the firefighter's role is limited to operating the tools that facilitate the rescue of the trapped occupant. While operating tools is important, the firefighter's evaluation and decision-making skills can have a great impact on the overall outcome of the incident. The firefighter's responsibilities can include the following:

- Identifying hazards around the scene, particularly under the vehicle
- Identifying the number of victims involved in the crash
- Contacting victims and determining injuries
- Determining degree of entrapment
- Identifying the best tool for the job
- Recognizing when a selected tactic is not working as anticipated
- Estimating the time required to perform the extrication

Most of the tasks a firefighter performs are individual tasks, such as operating the spreaders, cutting with an air chisel, etc., but firefighters are also part of a team with a common goal. This means that there are going to be situations when the company officer directs one firefighter to back out of the primary operating position and directs another into the area with a different tool. This action is often based on the company officer's observation of progress and the apparent need for a change in the approach being used.

The smart firefighter will back out and let another crew member take over, recognizing that the change is not because of any type of personal failure, but more a case of the bent metal not cooperating with the team. After backing out of the operating position, the firefighter can rest for a moment, reevaluate the task at hand, and be ready to return to work when directed by the company officer.

Summary

Successful operations at the crash scene are the product of strong leadership, capable tool handling, and teamwork. Aggressiveness must be tempered by planning, and individual victories must be considered secondary to the success of the overall operation. Clarity of thought and purposeful action, especially in the most challenging situations, will lead to the right solutions and the best results.

Appendix: Resource Utilization

Case Study: Overturned Tractor Trailer at the Edge of the Everglades

This incident provides an example of the wide variety of resources the company officer may have to consider utilizing at a serious crash scene. Starting with the scene size-up, the officer should predict the type and quantity of resources needed to perform the extrication. In this crash, the tractor trailer rolled over in a remote section of the East Everglades. When time and distance are factors, the company officer needs to request resources as soon as it becomes apparent that they are needed. Should the extrication be completed prior to the arrival of additional units, they can be returned to service.

Scene 1: The crash scene was blocked by trees that were cut by a unit that carried a chain saw. Creating good access to the vehicle and the victim at the beginning of an incident may save time later.

Scene 2: The patient met the Trauma Transport Criteria for transportation to the Ryder Trauma Center, a Level 1 center. The remote location of the crash required the dispatch of a helicopter for transportation. During the size-up, it was determined that the extrication of the victim was going to be complicated and that surgical intervention may be needed. The department's Assistant Medical Director, a surgeon at the trauma center, was transported to the scene by a second helicopter.

Scene 3: When a surgeon comes to the scene, a "Doc box" containing surgical instruments and supplies is brought to the scene by helicopter or another specialized unit.

Scene 4: Ultimately the victim was extricated without surgical intervention.

Scene 5: A Stokes basket was selected as the best choice for victim packaging.

Scene 6: The terrain was rough, making it difficult to carry the Stokes. Instead of a few individuals carrying the basket and running the risk of falling, a line of firefighters passed the basket toward the street and the waiting helicopter.

Vehicle Extrication: A Practical Guide

Scene 1

Scene 2

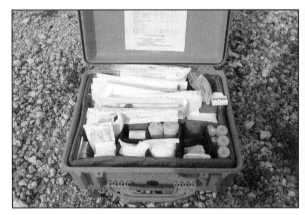

Scene 3

Scene 4

Scene 5

Scene 6

·2·

Tools, Equipment, and Apparatus

INTRODUCTION

This chapter addresses general operating procedures and maintenance for many tools, along with guidelines for cribbing. The purpose of this chapter is to help the reader learn about the application of basic extrication tools while avoiding the specific operational guidelines contained in an owner's manual. Updates and modifications of specific equipment and applications would make any attempt to cover brand-specific operating instructions inadequate and inappropriate. An owner's manual can provide a wealth of knowledge for new equipment.

PERSONAL PROTECTIVE EQUIPMENT

The personal protective apparel worn by crew members at crash scenes varies greatly and is a result of personal experience, comfort, organization policies, and a host of standards. The easy way to approach the subject of personal protective equipment (PPE) is to simply say that all personnel working at a scene should be provided with a complete ensemble that protects the crew member from the hazards posed by fire and biohazards, provides resistance to cuts and abrasions, and has good reflective properties when worn in low light conditions. Add to these requirements the need to comply with all federal, state, and local safety standards—along with standards created by other organizations, and it may become difficult to put together PPE that is reasonable to work in at a crash scene.

Casual observation of crash scene photographs and video images reveal that crews across the nation wear different types of protective gear. Some wear no protective gear at extrication scenes, while others may wear a complete ensemble designed for structural firefighting. Why is there such a variation of PPE when everyone is doing the same job? When equipped with structural firefighting gloves, why do some experienced and talented crew members pull a pair of lightweight leather gloves out of their pockets as they prepare to go to work at a crash scene? When equipped with goggles, why are crew members seen working without any eye protection?

Honest answers to these questions can only be obtained by asking those crew members who work in the gear on a daily basis about their choices. What may seem to provide optimal protection in an equipment catalog may not be the most functional piece of equipment at the crash scene. If the equipment that crew members prefer to wear doesn't meet established standards, steps can be taken to understand why there is a gap between the written standard and the practical standard. By identifying both the desirable and undesirable qualities of extrication PPE, perhaps improvements in gear and standards may evolve, providing crews with gear that is good to work in and compliant with standards.

Coats, pants, and boots

Structural firefighting coats and pants provide good protection from injury because of the multiple layers of materials used in the construction of these garments. The thickness and cut-resistance of the gear allows firefighters to work close to jagged metal, knowing that there is little chance of the metal penetrating all layers of the gear. The bulkiness and restriction of movement common in structural firefighting gear has resulted in experimentation with lighter weight materials, like coveralls and specialty extrication suits. While these specialty suits may be more comfortable to work in, they don't provide the protection from cuts and punctures that structural gear provides. Additionally, carrying extra sets of gear in a crowded cab may be impractical. Of course, crew members should wear steel-toe boots, which provide good protection from injuries to the feet.

Gloves

Standard structural firefighting gloves provide good protection for the hands and wrists when performing an extrication but may make the operation of small devices, like couplings and come-alongs, difficult. Many crew members experiment with lightweight leather work gloves, only to learn that they don't provide adequate cut resistance and coverage for the wrist area. Unlike firefighting gloves, many lightweight leather gloves lack the wristlets that prevent glass and other sharp objects from entering the glove. Specialty gloves have been introduced into the marketplace and are designed for special jobs like extrication. Gloves should be carefully examined to determine if they will actually hold up in the conditions encountered on crash scenes. After wearing different types of gloves, many individuals gravitate back towards a good-fitting, flexible, firefighting glove as the primary source of hand and wrist protection.

Head, face, and eye protection

The structural firefighting helmet has always been the first choice of firefighters working at crash scenes because of its built-in eye protection. The common face visor provides a good amount of face protection but only limited eye protection. As equipment and apparel have evolved to meet the needs of the fire service in the technical rescue field, many crew members are finding the smaller helmets used in structural-collapse and confined-space work to be useful in the extrication setting. The smaller profile allows crew members to work in tighter spaces without interference from the helmet brim or visor; when worn with goggles, it provides good head and eye protection but very limited face protection.

Eye protection has always been a concern when working with extrication tools at both emergency and non-emergency scenes. In the non-emergency setting, such as checking out and maintaining equipment, hoses are still under pressure and can cause powerful hydraulic leaks that often find a way to the operator's face. At the emergency scene, hoses operating at high pressure, cables under tension, and flying objects all pose a danger to crew members. The chance of injury associated with these conditions can be reduced with appropriate eye and face protection. As stated earlier, the ideal standards-compliant PPE may not be the most practical PPE for everyone.

For the past 25 to 30 years, the polycarbonate visor has provided some protection to the face and eyes, but not complete, protection. Visors are often kept in the *up* position to facilitate verbal communications, and to provide a clear view for the wearer. While new visors that are just out of the package provide good visibility, the visors of crew members working in a busy fire area will be stained, scratched, and distorted—making their use impractical. These crew members are more likely to keep the visors in the *up* position when they're unable to see their work through a damaged visor, and this exposes their eyes to various hazards.

Until recently, a good solution to the problem was the use of safety glasses in combination with the visor. The safety glasses provide good, but not complete, eye protection, while the visor provides some face and eye protection. In recent years, manufacturers have produced safety glasses that fit well and can be worn for long periods of time without causing a headache. These glasses are small enough to be stored in a coat pocket or in the apparatus.

Though it may seem trivial, the design and appearance of modern safety glasses is far superior to the inexpensive, old-fashioned safety glasses, making them more likely to be worn by crew members. The wraparound lens design provides good peripheral visibility—an important feature when watching for traffic at the crash scene and working in tight areas. Many experienced firefighters have found that combining safety glasses with a visor is the most effective way to provide good visibility and protection for the eyes and face.

Recent changes in eye protection standards have resulted in goggles being designated as the approved source of primary eye protection.

Goggles provide a better seal between the face and the goggles, which offers more protection from sprays and flying objects. With this improved sealing comes an unfortunate side effect: fogging. Goggles may work well when used indoors, in factories and labs, and under controlled temperature and humidity conditions, but when worn outdoors in the sun, condensation and fogging can occur. This can obscure the wearer's vision.

When these conditions are encountered, the wearer is torn between pulling the goggles off, thus having no protection or trying to see through the obscured lens. This problem may not occur in all areas of the country but is known to create problems in sunny south Florida. To fix this problem, some manufacturers have added vents to their goggles, which can help but don't solve the problem completely. During long operations, goggles not equipped with vents have been known to catch so much perspiration that it pools at the bottom of the goggles, sloshing from side to side as the wearer moves his or her head. Another problem that is encountered by some individuals while wearing goggles is the inability to breathe properly—commonly caused by a deviated nasal septum or broken nose. This problem is due to the design of the nosepiece and is not a problem with all brands of goggles.

The issue of compliance

Given the advantages and drawbacks of the different types of protective gear available, a gap may appear between the written standards created by some organizations and the practical standard created by crews that wear the gear on a daily basis. The question then becomes, *which standard should be used?* In a perfect world, there wouldn't be a gap between standards and practicality, thus there would be no problem. But in the real world, standards, rules, and contracts all have an impact on the safety of crews operating in the street. By careful examination of the standards, especially the disclaimers and notices sections, policymakers may find the solution to the problem.

In these disclaimers, the standards organizations may write that they don't test or evaluate the equipment; neither do they verify the soundness of the standards. If the standards organizations aren't responsible for the standards, then the responsibility for selection and use of the equipment would seem to fall on the end users. While this section of the book shouldn't be considered any kind of legal opinion, it should serve as a prompt for fire department administrators to carefully read the standards before adopting them and requiring compliance by its members.

TOOLS & EQUIPMENT

Cribbing

When the gap between the ground and a vehicle needs to be filled for support, cribbing can be used to fill the space. Cribbing is usually performed by stacking and arranging wood blocks to help support a load. In the vehicle extrication setting, 2x4s, 4x4s, and 6x6s are commonly used; they are cut into 18-in. or 24-in. lengths and used as wedges. The preferred woods for cribbing are Douglas fir or southern yellow pine because of their strength and crush characteristics. Pressure-treated lumber, while commonly used, is not considered a good choice because of its tendency to crack and split. There are also some manufactured plastic cribbing products available.

Tools, Equipment, and Apparatus

Cribbing can be used to limit movement in a variety of situations, including the following:

- A vehicle is in its normal position (four tires on the ground) but moves up and down as extrication procedures are being performed.

- A vehicle is on its side and is at risk of rolling over onto its roof or wheels.

- A vehicle is lying on its roof, allowing the vehicle to rock.

- A vehicle's roof posts are at risk of collapsing and need support.

Cribbing can be configured in a few different ways, but the most common is the box crib or its variations (see Fig. 2–1).

The most basic configuration consists of two pieces of cribbing laid parallel to each other, with two pieces laid perpendicular to the first two. This perpendicular or cross-tie configuration creates a wide, sound base, which is unlikely to fall apart when loaded from above. When support is needed in the center of the box crib, additional pieces of cribbing can be added to create a more solid platform. When a lot of downward force is anticipated, it's best to build a solid box crib that transfers the entire load through the wood to the ground without any open gaps. In some situations, a standard box crib won't fit in the available space, requiring the shape of the box crib to be modified. While circumstances may call for a modified box crib, it should be noted that modifications aren't as stable as a standard box crib; modified box cribs are classified with a lower capacity than the standard box crib configuration.

If a large amount of cribbing is needed quickly in unusual circumstances, long 4x4s can be cut on site to create cribbing blocks and wedges. A large circular saw can be used to cut blocks, or a chain saw can be used to cut blocks and wedges (see Fig. 2–2).

Cribbing is also used as a foundation or platform to support loads that are added during the extrication. Situations benefiting from cribbing to support loads include the following:

- When dash lift or dash roll-up procedures are performed, cribbing is placed between the ground and the underside of the vehicle. This provides the strength needed to resist the force created by the tools. By filling this gap, the portion of the vehicle that is being used as a push-off point (or base) is less likely to bend during the operation.

- When lifting procedures are being performed, the effectiveness of the lifting tool can be improved by closing the gap between the ground and the lifting point on the vehicle. This may be accomplished with the use of a single 4x4, or may require a box crib, depending on the tool and the circumstances. The cribbing used to build the base is essentially artificially raising the ground, allowing the tool to be used more effectively.

- When pulling may cause the surface of an object to collapse, cribbing between the surface of the vehicle and the tool or chains can spread the force exerted over a large area, decreasing the chance of the object collapsing.

Chains

In the extrication setting, chains are usually attached to a winch, come-along, or hydraulic rescue tool to perform some type of pulling operation. Chains are used as accessories with these tools for the following reasons:

- They are tough and difficult to damage; they can come in contact with bent sheet metal with little chance of damage.

- Chain sets can be assembled in a variety of ways to meet the needs of the user.

2x2 box crib

3x3 box crib

2x2 box crib with a solid top

Solid box crib

Modified box crib

Modified box crib

Fig. 2–1 Box cribbing basic configurations

Tools, Equipment, and Apparatus

- With reasonable care, chains have a long life span.

- Chains don't store potential energy.

- Chains are simple to understand.

Understanding the proper way to select and use chains is important in obtaining the desired results. Before selecting and using a chain, it's a good idea to understand some chain basics.

A work stand can be built quickly from cribbing.

The bar of the chainsaw is set up to cut from one side of the 4x4 to the other.

The end of the chainsaw bar must project beyond the edge of the wood to complete the cut.

The near side of the bar is positioned at the corner of the 4x4.

During the cut, the saw bar should be kept vertical to keep the cut straight.

After the first wedge is cut, a straight cut is made, creating a second wedge.

Fig. 2–2 Cutting wedges with a chain saw when more wedges are needed in a hurry

Chain construction and grading

Chains are made of metals that make them strong and durable. The metals that go into making a chain determine the strength and subsequent rating of a chain. There are all kinds of chain shapes and strengths, but the two that are most commonly used in the fire service are grade 70 and grade 80 chains. The grading system for chains is rather simple, and, in most cases, the higher the number the stronger the chain. For example, a chain with a nominal size of 3/8 of an inch and classification of grade 30 may have a capacity of 2650 lbs. A grade 70 chain may have a capacity of 6600 lbs, and a grade 80 chain may have a capacity of 7100 lbs. Grade 70 and grade 80 chains are preferred for the fire service because of the relationship of the weight of the chain to the strength of the chain. It makes sense to carry a lighter, grade 80 chain if it's just as strong as a larger, heavier, and lower-grade chain. Even though the numbers cited in these examples come from a manufacturer's products manual, they should be viewed as illustrative only. The actual specifications and design of a chain sling should be developed by a qualified chain representative.

Grade 70 chain is also known as transport/binder chain and is commonly used by truckers and loggers for securing loads. Grade 70 chain is made of heat-treated carbon steel. This type of chain can be identified by the embossing code on the links of G70, 7, 70, G7, M7, and C7. The number 7 in the embossing helps to identify the chain as a grade 70 chain. This chain may have a bright yellow/gold finish.

Grade 80 chain is an alloy chain that can be used for overhead lifting in an industrial setting. Grade 80 chains are made of heat-treated alloy steel. These chains can be identified by examining the links for the embossing of G80, HA800, A8A, 8, TC8, CA8, and G80. The number 8 in the embossing helps to identify the chain as grade 80.

Other components of a chain set

To be used properly, the chain needs additional components such as master links, hooks, and identification tags.

Master link. The master link is the connecting point between the chain and the pulling device. On a winch or come-along, the master link is attached to the pulling device's sling hook. By using a master link, damage to the chain is avoided by providing a properly designed pulling point. If more than one chain or leg is desired in the sling set, the other chains are also attached to the master link. Master links can be oblong, circular, or pear-shaped. For this reason, the oblong-shaped master link in the typical fire service chain sling is also known simply as an oblong ring.

Hooks. There are two types of hooks commonly used in fire service applications: sling hooks and grab hooks. Sling hooks can be used to attach a chain sling to a built-in lifting point, as seen frequently on military equipment. Sling hooks can also be used for wrapping around an object and hooking back into the chain in a choking manner. A desirable additional attachment to a sling hook is a latch. The latch is a spring-loaded clip located over the opening of the

Tools, Equipment, and Apparatus

hook that prevents the hook from disconnecting during the set-up phase of an operation. When the sling is being set up, it's important that there is no load exerted on the inside of the latch, as the latch is not designed to carry any load. In the extrication setting, the sling hook is more versatile than the grab hook—making it the preferred hook at the end of the chain opposite the master link.

The front of the ID tag with serial number, size, and length of chain.

The rear of the ID tag provides the capacity and chain set configuration.

The master link, also known as an oblong ring, with a chain shortener assembly.

This grade 80 chain can be identified by the number 8 in the code.

This sling hook is at the end of the chain and is equipped with a latch, a useful accessory.

Grab hooks are used on chain shorteners but aren't typically attached to the end of the chain.

Fig. 2–3 Chain set components

Grab hooks, on the other hand, are designed to grip an individual link on a chain set, creating a loop of chain that is fixed in length. The grab hook is the type of hook needed for chain shortener assemblies or for extension chains. By grabbing an individual link, the chain length remains fixed and doesn't choke down. This is a helpful feature when two chains need to be connected.

Identification tag. When purchasing a chain from a quality chain distributor, the inclusion of an identification tag provides the users with the confidence that the chain was properly constructed and rated. Metal identification tags are usually attached to the master link and contain some (or all) of the following information, which is useful for tracking and identifying chains and their capabilities.

- Size of the chain
- Length or reach
- Working load limit at 90°
- Serial number
- Manufacturer's name
- Grade of chain
- Sling type (configuration)
- Date of assembly

Typical chain configurations

Come-along chains. The most common chain sling configuration for use with come-alongs consists of the following parts:

- Master link—attaches to the hook on the come-along.
- Chain shortener assembly—attached to the master link. It allows the chain to be shortened after the slip hook end of the chain is set up.
- Twelve ft of grade 80 chain—12 ft is a good length for day-to-day use. The size and capacity of the chain depends on the capacity of the pulling device.
- Sling hook with latch—attaches at the opposite end of the chain from where the master link is connected.

Hydraulic spreader chains. There are a couple of configurations available for hydraulic tools. The method of attachment is the most significant variable. Following are examples.

- Sling hook attached to the end of chain—in this set up, grab hook shackles are attached to the spreader arms, or both ends of a ram. One sling hook is attached to an object that serves as an anchor point, and the other is connected to the object to be pulled. Once the sling hooks are attached, the chains are brought back to the tool and attached to the grab hook shackles.
- Sling or grab hook at one end, shackle at the other end of a length of chain—the shackle is designed to attach directly to the attachment holes on the spreader arms or the ends of the rams.
- Sling or grab hook at one end, shackle at the other end, with a chain shortener device—this set up allows the slack to be taken out of the chain after the chains are attached at all four ends.
- Grab hooks at both ends of the chain—this configuration is used as a chain extension. The grab hooks are used to attach to other chains.

Tools, Equipment, and Apparatus

Following are a few points that should be remembered when using chains.

- The weak link—in any pulling system, one component in the system has to be the weakest. In a typical come-along system, consisting of two chain slings and a come-along, the weakest part of the system should be the handle on the come-along. The chain or cable on the come-along should continue to perform without failure long after the handle fails. The reason for this is simple: it's much safer to have the

The simplest way to attach the chain to an object is to use the sling hook.

The length is then adjusted using the chain shortener.

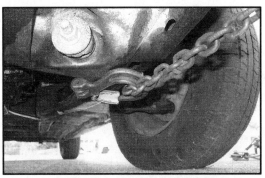

In this case, the chain was wrapped around the chassis then choked with the sling hook.

The chain shortener is also used when the sling hook is used to choke.

If choking isn't the best option, the chain can be wrapped around the object.

Then brought back and attached to the grab hook on the chain shortener assembly.

Fig. 2–4 Basic chain procedures—choices for attaching and adjusting chain lengths

handle fail when under a 2-ton load than to have the cable or chain fail. When the handle fails, it's simply going to bend. If the cable or chains were to fail, the force of the entire system would be released in an uncontrolled, dangerous manner. The key is to use chains that exceed the capacity of the tools to which they are attached.

- Don't rely on chains that have uncertain ratings or history. That mystery chain found in the back storeroom under the workbench at the firehouse isn't the chain to use at a rescue scene.

- Hooks shouldn't be tip loaded. Hooks are designed so that the load is located deep in the hook. If a hook is loaded on or near the tip, additional leverage is created, which can cause the hook to open up and fail. For this reason, the tip of a hook should never be inserted into the middle of a chain link. It's not good for the chain, and it's not good for the hook.

- Even heavy chains can be overloaded to their breaking point. While chains don't store potential energy, the object they are attached to may be storing energy that could be released suddenly and violently if a chain were to break.

- Chains that are rated for use on a come-along are unlikely to be rated for use with hydraulic tools. It's best to use the matched chain slings provided by the hydraulic rescue tool manufacturer for use on the hydraulic rescue tools. To reduce the chance of confusion, the chains can be clearly marked indicating which tools should be used with the particular chain set.

Come-alongs

Come-alongs are simple, old-technology tools that can be used for numerous types of pulling procedures at the scene of a crash. The term *come-along* is the generally accepted substitute for the proper term *cable hoist*. A typical come-along has a pulling capacity of 2 to 3 tons and is hand operated. The tool develops its power by mechanical advantage, which is created by the handle and the cable pulley. There are three similar devices used in the fire service.

- Cable hoist—more frequently referred to as a come-along. With this tool, a steel cable wraps around the drum when the come-along is operated.

- Strap hoist—similar to a cable hoist, but uses a strap to wrap around the drum. While the strap hoist is easier to use than a cable hoist, the strap is more susceptible to being cut in extrication operations.

- Chain hoist—a fairly heavy tool that uses a chain instead of a strap or cable. In addition to the weight factor, the length of chain built into the tool limits the pulling distance that can be achieved.

All three of these tools provide the ability to pull, but the cable hoist provides greater overall utility of the three.

Come-along components

There are only a few operating controls on a come-along, but training and familiarity are necessary to achieve the best results, especially in critical situations. With the system under a heavy load, untrained or inexperienced operators are often reluctant to operate any of the controls other the cranking handle because of concern of suddenly

Tools, Equipment, and Apparatus

releasing the load. With adequate training and practice, a come-along can be used very effectively in a wide variety of pulling situations.

All come-alongs consist of the following parts.

- Body—the body of the come-along is cast metal component to which the other parts are attached.

- Drum—the drum is located in the body of the come-along; this is where the cable is wrapped.

- Cable—when a come-along is operated, the cable is wrapped around the drum. The cable is attached to the drum on one end, and to a hook on the other end. As the come-along is operated, the cable wraps around the drum and pulls the object closer to the come-along.

- Handle—the handle has two important roles in the operation of a come-along. First, it helps provide mechanical advantage when the cable is wrapped onto the drum. Second, the handle serves as an indicator that the maximum capacity of the come-along has been met. When the operator has exerted the force needed to meet the capacity of the come-along and then starts to exceed it, the handle will start to bend.

- Bale—the bale is where the handle attaches to the come-along, and where the direction-of-travel lever is located.

- Direction-of-travel lever—this lever controls whether cable is wound on or wound off the drum.

- Free-spool lever—operating this lever allows the drum to spin freely without engaging the ratcheting teeth.

- Hand wheel—the hand wheel is used to quickly wind slack cable onto the drum. Using the handle wheel provides virtually no mechanical advantage, so it isn't used once the come-along is placed under a load.

- Hooks—there are three hooks on a typical come-along. One hook is attached to the body of the come-along, another is attached to the end of the cable, and the third is part of the block. The hooks are used to attach the come-along to an anchor point and to the object that will be pulled.

- Block—to achieve a 2:1 mechanical advantage the cable passes through the block and is attached back to the come-along body. The block consists of an enclosed pulley and a hook. If a lot of cable is required in an operation, the come-along can be used in a 1:1 mechanical advantage. In this configuration the block is not used—it simply rides on the cable during the pulling operation.

- Rating plates or labels—these are important when selecting the chain sets to be used with the come-along. The capacity of the chain should exceed the capacity of the come-along. When a come-along system is built using the come-along and chain sets, the weakest part of the system should be the come-along handle.

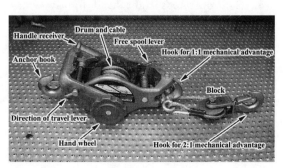

Fig. 2–5 Come-along parts

Basic come-along actions

Mastery of the following basic actions will provide the operator with the skills needed to operate the come-along quickly and confidently in the rescue setting.

Letting out cable without any load.

The come-along should be stored with the cable wound snugly on the drum but with a few inches of slack at the hook end of the cable. Follow the steps below when it's time to use the come-along for a pulling operation:

1. Connect the block hook (2:1 mechanical advantage) or the cable hook (1:1 mechanical advantage) to the object that is to be moved. If a chain set is used to wrap around the object, the hook on the cable or block should be attached to the master link of the chain set. If the object to be moved has a hoisting eye or other good, strong attachment point, the come-along hook can be attached directly to the object, eliminating the need for a chain. If there is any question about the strength of a direct attachment point, a chain should be used to wrap the object instead.

2. Flip the direction-of-travel lever to the position that allows the cable to pay out off the drum.

3. The operator's right hand lightly wraps around the hand wheel, providing support and balance while at the same time providing a light friction that will prevent the cable from becoming loose on the drum, creating a *backlash* or *bird's nest*.

4. The operator's left hand supports the body of the come-along, while the left thumb depresses the free-spool lever.

5. With the free-spool lever continually depressed and light pressure on the hand wheel, the operator backs away from the object.

6. Once the desired amount of cable has been removed from the drum, the free-spool lever is released. The come-along is then ready to be hooked to the anchor point chain.

There are a few problems that commonly occur when letting out cable.

Problem #1: With the direction-of-travel lever in the correct position and the free-spool depressed, the cable will not pay out and the drum won't rotate even a small distance.

Cause: Even though the direction-of-travel lever has been flipped to the pay out position, the pawl hasn't disengaged from the teeth on the drum. This is caused by friction between the parts, and a slight amount of tension, which is holding the parts against one another. The come-along is a simple, low-tech device, and this type of problem occurs on occasion.

Solution: With the direction-of-travel lever in the correct position to pay out cable, the hand wheel is rotated slightly in the winding-on direction to release the pressure between the pawl and the tooth so they can disengage.

Problem #2: When backing away from the object that is to be pulled, the cable on the drum becomes loose and creates a tangle of loose cable commonly known as a bird's nest.

Cause: As the cable is taken off the drum, the spinning drum develops some momentum. If the cable is not kept taut as it pays off the drum, the cable can unspool around the drum, creating the bird's nest.

Tools, Equipment, and Apparatus

Solution: The best solution is to apply light pressure or friction to the hand wheel with the operator's right hand as the cable pays out. By providing light friction the drum will stop turning immediately when the cable stops paying out, keeping the cable wrapped snugly on the drum. Another solution that can be combined with the light pressure is to always have the come-along operator back away from the object to be pulled when paying out cable. By backing away from the object instead being stationary and having the cable pulled to the object, the operator has better control on the cable tension as the cable is pulled out. When the operator backs away from the object to be pulled, control of the speed of pay out and the friction needed to control the cable is literally in one person's hands.

Problem #3: The newly purchased come-along develops bird's nests more frequently than the old one that was purchased 15 years ago.

Cause: When the come-along is new, the cable hasn't had a chance to conform to the circular shape and size of the come-along drum. Before the come-along was manufactured, the cable was probably stored on a much larger spool that allowed the cable to retain its natural, straight shape. Once loaded on the come-along drum the cable wants to straighten out again and will try to unwind if wrapped loosely on the drum. After using the come-along for a while, the cable will start to bend and adapt to the new shape. This break-in period will take longer than it did ten years ago because manufacturers are now using larger diameter cables, which take longer to adjust to their new shape.

Solution: Train and drill with the equipment. The more the equipment is used the quicker the cable will conform to its new shape, which results in a reduction in the number of birds' nests.

Taking up the slack cable. After the come-along has been attached between the object to be pulled and the anchor point, the slack in the cable should be wound onto the drum. Taking the slack out of the line is important for the following reasons:

- If there is any loose cable on the drum, it needs to be eliminated before the system is put under load. If the slack isn't removed, the inner slack windings of cable can become damaged when outer layers of cable are wound over them. When the cable is damaged in this manner, it's usually flattened or severely bent. Put in the simplest terms, the slack cable gets squashed.

- It allows the system rigging to be checked for correctness before a heavy load is put on the system.

- By taking the slack out of the come-along, the system is put under a light load, which usually allows the system to support its own weight without shifting or falling. This also allows the crew members to make final adjustments, add additional cribbing, and insert the handle into the come-along without the come-along flopping around.

There are a couple of problems that can occur when taking up slack.

Problem #4: When the bale is operated, the cable doesn't wind onto the spool.

Cause: The pawl isn't engaging the sprocket on the drum.

Solution: Flip the direction-of-travel lever from the payout position to the take-up position.

Vehicle Extrication: A Practical Guide

Problem #5: With the handle in place it's difficult to manipulate the come-along when trying to take up the slack. The come-along has a tendency to flip upside down causing the handle to be pointing toward the ground.

Cause: With no load on the system, all the chains and cable are loose. When the handle is inserted into the come-along the come-along becomes top heavy and spins on its longitudinal axis.

To clearly show the procedure, the scenario will depict a car being pulled to a new position.

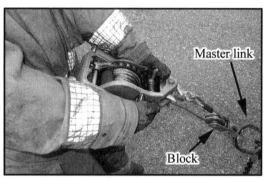

The block is attached to the master link of the object that will be moved.

To let out cable, the free spool lever is depressed while light tension is placed on the hand wheel.

The operator backs away from the object that is to be pulled.

The come-along reaches the master link of of the anchor chain.

The hook on the body of the come-along is then hooked to the master link.

Fig. 2–6 Basic come-along procedures—setting up to pull

Tools, Equipment, and Apparatus

Solution: For ease of set-up, the handle shouldn't be inserted into the come-along until there is slight tension on the system. It's best to make all the hook-ups and then take the slack out by using either the hand wheel or the bale. While they don't provide as much mechanical advantage as the handle, they are easier to use until the system has a slight load placed on it.

Problem #6: The controls seem to be operating backwards.

Cause: The come-along is rigged backwards or the come-along is upside down.

Solution: This is a problem that is usually corrected by developing a routine when setting up the come-along for a pulling operation. For example, if during training, the operator always cradles the hand wheel in the right hand, depresses the free-spool lever with the left thumb, and backs away from the object that is to be pulled, then the operator will immediately sense when the come-along is not in the correct position when being set up. The operator will sense that things are not where they should be and can quickly identify the problem and correct it. This routine will be extremely helpful in the unusual setting when the operator can't use the normal routine. By having a good understanding of the normal feel of where controls should be, the operator will be better able to adapt to the unusual situation that may require the come-along to be rigged upside down, etc.

Operating the come-along with no slack and under load.

With the slack cable taken out of the system, the come-along is ready to start the pulling operation. Before any additional pressure is put on the system, the components should be quickly checked for any rigging problems that might be present. The most thorough and efficient way to check the system is for the person who rigged the system to check the system. For example, the crew member who rigged a door that is to be pulled should check the chain that wrapped the door and all the connection points back to the come-along. The crew member who made the anchor hookup should check from the anchor point to the come-along. To check the system, crew members should perform the following steps.

- Look at the hooks to make sure they won't snag other objects during the pulling process.

- Make sure the chain is properly seated in the hook if using grab hooks.

- Make sure the chain is in the bowl, or thick part of the hook—not hooked on the tip of the hook through the middle of a link if using sling hooks.

- Untwist any cables that have become twisted on themselves.

- Check the cable hook to ensure that the cable comes off the hook in a straight line and isn't looped around the hook or itself.

After checking the system and correcting any problems that may have been present, the come-along can be operated. The handle is inserted into the come-along and locked into place. The most common devices for locking the handle in place are spring clips or locking screws. These locking devices prevent the handle from sliding out of the attachment point when it is being operated.

Most come-alongs allow the handle to be attached from two directions. The normal or typical position is to insert the handle from the top of the come-along. When using this method, the handle is inserted into the bale with the come-along in the normal operating position: bale up, hand wheel on right side, free-spool lever on the left side. In some situations, depending on operating clearances and obstructions, it may be beneficial to insert the handle from the underside of the come-along. In addition to being awkward, the alternative position provides a little less mechanical advantage, because the full length of the handle isn't being used. Having this option, while not used often, can be a real plus when operating in tight or unusual situations and when the full length of the handle would make it difficult to operate the come-along.

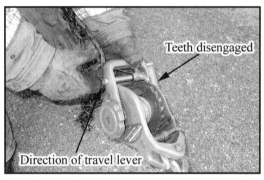

To start the pull, the direction of the travel lever is moved from the disengaged...

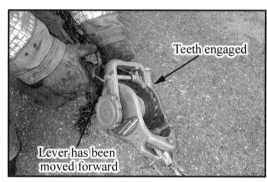

...to the engaged postion.

The hand wheel is used to take up the slack quickly.

The bale can also be used to take up slack cable.

With the slack taken out, the handle is inserted and used for greater leverage.

For maximum leverage, the handle should be gripped at the end.

Fig. 2–7 Basic come-along procedures—making the pull

Tools, Equipment, and Apparatus

If the come-along is going to be used in the 1:1 mechanical advantage configuration with almost all the cable being used for the pull, it's a good idea to place a light object such as a bunker coat or floor mat on the cable. If for some reason the system should fail and come apart while under load, the object laid on the cable will cause the cable to fall to the ground, thereby reducing the chance of the cable springing back and injuring one of the crew members. Usually, this procedure is not used when the cable is used in the 2:1 mechanical advantage configuration because the cable is usually short and is operating in a congested work area.

With the system checked and ready to be operated, the come-along operator should grip the handle at the end to obtain maximum leverage. At first, the handle can be operated relatively easily, but as the maximum capacity of the come-along is approached, good body mechanics are needed to reduce the chance of injury and to obtain maximum performance from the tool. This means the operator should use the muscles of the arms, shoulders, and legs to operate the come-along. Use of the operator's back for leverage can lead to injury and should be avoided.

As the come-along approaches its capacity rating, the operator will be working hard to push the handle. At this point the operator should watch for any bending of the handle, indicating the capacity of the come-along has been met. If the handle starts to bend, other methods of performing the procedure will have to be developed. It's important to remember that in addition to providing leverage to operate the come-along, the handle is a safety device that shouldn't be circumvented. Crew members should resist the urge to remove the handle and replace it with a longer or stronger object such as a piece of pipe. Use of pipes or *cheater bars* can result in another part of the come-along failing with bad results.

There are a couple of problems that can occur when operating the come-along under load.

Problem #7: When the come-along is being used in the 2:1 mechanical advantage mode (utilizing the block and doubling the cable hook back to the come-along body), the cable *occasionally* comes off the pulley sideways and gets jammed.

Cause: When the system was being set up, the cable and block were not kept in a straight line. Usually, a piece of cribbing or other obstruction causes the cable to come off the pulley wheel and side load the pulley block. As more tension is put on the system, the cable may become wedged between the pulley and the body of the block.

Solution: This problem can be fixed by releasing the tension on the system and pulling the cable out. Unfortunately, the problem often isn't recognized until the cable is jammed so badly that the come-along stops operating. In this case, it is usually necessary to disassemble the pulley block to release the cable.

Problem #8: When the come-along is being used in the 2:1 mechanical advantage mode (utilizing the pulley block and doubling the cable hook back to the come-along body), the cable *frequently* comes off the pulley and gets jammed.

Cause: The cause of this problem may be poor design or fabrication of the come-along. The professional grade come-alongs are designed so the cable doesn't easily come off the pulley and get jammed.

Solution: Purchase quality, professional-grade come-alongs, and carefully inspect and train with the equipment.

Releasing cable tension when under load. Letting cable out when the system is under load is the procedure that causes the most anxiety for inexperienced users of a come-along. Nobody wants to have a system that has two tons of load on it unload suddenly because of operational error. The simplest way to remember how to take a load off the system is to look at how the load was put on the system.

- If the load was put on the come-along by hand, using the hand wheel, the load usually can be taken off using the free-spool lever while providing light friction on the hand wheel with the right hand.
- If the load was put on the system using the handle for leverage, the load should be taken off the come-along using the handle.
- If the load was put on the system using the bale and applying only light pressure, it is probably safe to use the free-spool lever.
- If the load was put on the system using the bale and applying a lot of pressure, it may be best to use the handle to take the load off to maintain good control.

The procedure for using the free-spool lever to let out cable was covered earlier and it is fairly simple. The important point to remember is to avoid creating a bird's nest. Removing a heavy load from a come-along is also simple and only requires the following steps:

1. Insert and lock the handle into the come-along if it isn't already in place.
2. Flip the direction-of-travel lever to the payout position.
3. Push the handle forward firmly until a single clicking sound is heard. Maintain a firm grip on the handle, as this step creates some rearward force against the operator's hand that is holding the handle.
4. Allow the handle to travel rearward as the operator feels some rearward pressure against the hand with the handle in it until another single clicking sound is heard.
5. Repeat the forward and rearward movement until the desired amount of cable is released from the drum.

When the come-along is under load, there is no reason to operate or touch the free-spool lever, as this is being done automatically by the mechanism of the come-along as the handle is being operated forward, then rearward. Once the load has been removed from the come-along, and the cable is no longer under tension, the handle will stop paying out line, and the free-spool lever must be used to release more line.

There are only a couple of problems that can occur when paying out cable when the come-along is under a load.

Problem #9: When there is a very heavy load on the come-along it seems like the handle won't move all the way forward to release the load.

Cause: As the load increases so does the amount of force required to release the load using the handle.

Solution: Use more effort to move the come-along handle forward. While doing this, it is important to maintain a good grip on the handle as there will be force exerted rearward.

Problem #10: With a load on the come-along, the handle moves forward and rearward but doesn't release the load. The cable simply moves back and forth, but no progress is made.

Cause: The pawl on the come-along is engaging the teeth on the drum, but isn't moved

Tools, Equipment, and Apparatus

rearward far enough to disengage and reengage the next tooth.

Solution: As the come-along handle is moved forward a single click sound will be heard. As the handle is moved rearward the operator must bring the handle rearward enough to allow the pawl to disengage from one tooth to engage the next tooth. When releasing the load on the come-along the operator should hear a very distinctive click-click sound. Both clicks must be heard for the come-along load to be released one tooth at a time.

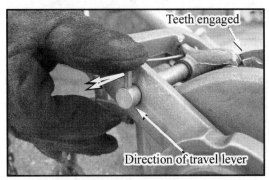

To take the load off a come-along, the direction of travel lever is moved rearward.

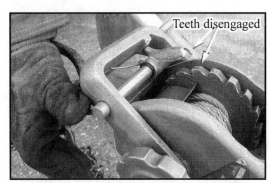

With the lever moved rearward, the teeth are disengaged.

While maintaining a good grip, the come-along handle is moved forward until a click is heard.

When the pawl engages the spring, the free spool lever will release one tooth at a time.

The pawl has engaged the spring.

After the first click is heard, the handle is allowed to move rearward until a second click is heard.

Fig. 2–8 Basic come-along procedures—safely releasing a load while under tension

Vehicle Extrication: A Practical Guide

For maximum leverage and pulling power, the handle is normally used on the left side.

In situations where space is limited, the handle can be reversed, taking up less space.

When more cable is needed, the cable hook can be disconnected and used in 1:1 ma.

When used in 1:1 ma, the block is simply slid out of the way.

This 3-ton Lugall come-along has no hand wheel. The cable still pays off smoothly without it.

For the leverage required to operate the 3-ton come-along, a telescoping handle is used.

Fig. 2–9 Come-along options and features

Air (pneumatic) tool systems

Air-operated tools have long played a role in vehicle extrication but have usually been considered a secondary, or back-up tool system. Prior to the 1970s, when engine-driven hydraulic tools came into widespread use, air tools played a greater role when they were combined with basic hand tools and manually operated hydraulic tools. Ironically, as the use of self-contained breathing apparatus (SCBA) became widespread, along with a readily available air source, the use of pneumatic tools continued to drop off. There are several reasons why some pneumatic tools have faded away.

- Engine-driven hydraulic tools were new and provided a tremendous amount of power for use at crash scenes. The new hydraulic tool systems could free victims in the vast majority of crashes, making the pneumatic tools mistakenly appear obsolete.

- The most common air tool, the air chisel, was a low-power tool, with a limited selection of bits. Commonly referred to as *muffler guns*, these tools were usually considered capable of cutting lightweight sheet metal—not a common problem on the crash scene.

- Air chisel operators require skill and practice to become effective when using the tools. In the hands of a skilled operator, the air chisel can work wonders, while in the hands of the inexperienced operator it can be a big time-waster.

- Tools like impact wrenches are used very infrequently, making them seem unimportant.

As the fire service's scope of responsibilities have expanded in the areas of extrication, confined space, and trench rescue operations, air tools have reemerged as an important tool system. As the need for air tools has increased, so have the quality, design, and functionality of the tools—making them an important part of the tool inventory. Following are the most common air tools used for vehicle extrication:

- Air chisels.
- High-pressure air bags.
- Impact wrenches.
- Die grinders/whizzers.

With a few adjustments, all of these tools can operate off of the same basic system, making them easy to set up and operate.

The air tool system overview

The air tool system consists of the following components:

- Stored air source—the cylinder, tank or compressor that provides the air needed to operate the tool

- Regulator—reduces the high pressure of the stored air source to a lower operating pressure needed for the tool

- Supply hose—transfers the air from the regulator to the tool

- Tools

Air sources

SCBA cylinder. Most fire departments use SCBAs, making the SCBA cylinder one of the most common sources of stored air for operating tools. These cylinders are plentiful, rugged, and relatively easy to handle and transport. The one drawback to the cylinders is their limited storage capacity. This capacity limitation becomes important when using tools that use a large amount of air, such as high-pressure air chisels and the largest high-pressure air bags. For example, air consumption in high-pressure air chisels can reach 15 to 20 cubic ft per minute (cfm), and a 70–ton high-pressure air bag can deplete an entire air cylinder with one inflation. The tools that require a lot of air to operate can deplete an air cylinder quickly, creating logistical and tactical challenges. Given no other air source, SCBA cylinders can be used with high-volume tools, but it will require a steady supply of cylinders and personnel assigned to swap them out as they become depleted. While there are drawbacks to using these cylinders, they are still one of the best sources for air on the rescue scene.

To increase the amount of air available with SCBA cylinders, carriers have been developed that joins two SCBA cylinders together, forming a mini-cascade system that increases the amount of air available before having to change out cylinders.

Large apparatus mounted cylinders. Some of the problems associated with the low volume (cubic ft) of SCBA cylinders can be eliminated when a fire apparatus is equipped with large air cylinders permanently mounted and connected to hose reels. These large cylinders contain a large amount of air at pressures of 3600–4500 pounds per square inch (psi), providing an adequate air supply for most extrication operations. While there are other capacities and pressures, these are the sizes typically found on a fire apparatus. The cylinders are rather heavy and awkward to move around, so the cylinders are mounted semi-permanently on the apparatus and attached to regulators and hose reels, or long coiled lengths of air hose.

Specialized apparatus. Departments that operate air trucks, air/light trucks, or scene support vehicles equipped with a compressor and cascade system, can provide a virtually unlimited supply of air to a crash scene. If these vehicles are outfitted to supply tool air, careful consideration should be given to all the applicable regulations and standards that are in place to prevent contamination of breathing air. In these vehicles it is best to work with the manufacturer of the apparatus, or the air system manufacturer, to ensure proper use of the equipment.

Set-ups to avoid. While at first glance these ideas may seem like possible solutions, if implemented, they could result in tragedy.

- Using large oxygen cylinders—oxygen cylinders should never be used as an air source. The fittings on air and oxygen equipment are different by design in an effort to avoid connecting the wrong components together. The ability to adapt is a positive trait in firefighters, but the idea of switching fittings to connect air tools to oxygen or any other gas should be prohibited.

- Using the fire apparatus air brake tanks—using the brake tanks on fire apparatus may at first seem like a good idea, but there are a few problems associated with this idea. First, and most importantly, if the air supply is drained off the air brake tanks, the pressure in the tanks will drop to a pressure that will prevent the apparatus from being moved. The need to move an apparatus after initial positioning at a crash scene doesn't occur very often, but an apparatus should not be immobilized by choice. Secondly, the air pressure in the air brake tanks will be in the 100-psi range, which is on the low side. Even if the pressure is adequate to run the desired tool the deliverable volume (cfm) is limited, and capacity of the storage tanks is rather

Tools, Equipment, and Apparatus

small. Additionally, the engine air compressor will not keep up with most high-volume demands. The air brake system should be considered no more than a back-up to the back-up air source.

Regulators

Regulators are necessary to reduce the high pressure of the air source, typically an air cylinder, to a lower-working pressure required by tools. To maximize the usefulness of an air tool system, it is important to understand how a regulator works. By understanding how a regulator works, important decisions can be made when specifying, using, and trouble-shooting the equipment. When specifying a regulator, the pressure and cubic foot per minute needs of the tools must be known to properly select a regulator that can keep up with the air demand. An understanding of how a regulator works is helpful when using other equipment requiring regulators such as air shores for trench rescue, and oxygen and acetylene regulators for cutting torches. To illustrate how a regulator works, each component of a typical regulator will be examined.

Inlet fitting. The inlet fitting connects the air cylinder to the regulator. It consists of a female threaded nut and a nipple that is designed to match the air source cylinder. Careful examination of this fitting provides information about the intended use of the regulator. First, there may be a designation stamped on the fitting with a series of letters and numbers. These letters and numbers indicate the approved use of the item as designated by the Compressed Gas Association (CGA). A common designation is CGA 347, which is a designation that indicates the fitting is designed for use on air cylinders with a pressure as high as 5500 psi. This designation is used to differentiate it from other applications, such as other gases and pressures. For example, if the stamped designation on the fitting reads "CGA 346," the fitting is designed for use on air cylinders with a pressure up to 3000 psi, eliminating its use on a 4500-psi cylinder.

Next is the length of the nipple that fits on the inside to air cylinder discharge. If the fitting is designed to be used with a 4500-psi cylinder, the nipple will be longer than those used with 2216 cylinders. The end of the nipple will be either a smooth, curved, brass surface, or a surface with a circular groove. If the nipple is fitted with a circular groove, the nipple is designed to use an O ring to create a seal between the nipple and the air cylinder discharge. If this O ring is missing, the regulator will not seal properly against the air cylinder unless an excessive amount of force is used.

Fittings that use O rings are designed to be hand tightened, while the smooth brass fittings are tightened with a wrench. For portable air tools that are stored on the apparatus disassembled, hand tightening is probably the way to go for quick set-up. The down side to the hand-tightened fittings is that when the system is being assembled at the crash scene, the O ring often comes up missing, making a seal difficult. Another clue that helps determine if the fitting is supposed to be wrench tightened or hand tightened is the presence or absence of a hand wheel. The hand wheel is supposed to give the appropriate amount of leverage to create a seal when tightened by hand. The absence of a hand wheel can mean that the fitting is supposed to be tightened with a wrench or that the hand wheel broke and wasn't replaced.

Regulator body. The body of the regulator contains the diaphragm and connects to all the other regulator parts. Markings on the body include the following:

- HP—indicates the high-pressure side of the diaphragm. This is the side that is connected to the air cylinder or other air source. The inlet pressure gauge and the air inlet coupling will be connected to the openings marked "HP."

- LP—indicates the low pressure or discharge side of the diaphragm. The discharge pressure gauge and hose coupling are attached there.

- Max. pressure in—indicates the maximum amount of pressure the regulator should receive from the air source. Pressures higher than indicated could result in regulator damage/failure.

Bonnet. Gauges can get beaten up when stored on the apparatus and when used on-scene. To reduce the damage inflicted on the gauges, a protective shield called a *bonnet* is placed around the gauges and attached to the regulator body. This is a good investment that will reduce later expense and downtime for repairs.

High-pressure gauge. This gauge measures the air pressure that is entering the regulator from the air source, so the reading on the gauge should match the pressure on the air cylinder gauge—give or take 100 lbs. There may be two systems of measurement on the gauge face: the U.S. Customary system, using psi and International System (SI) that uses kilopascals (kPa). Tradition dictates that the U. S. Customary be used in the fire service. After all, it's not very often a firefighter at a rescue seen yells out, "Hey Joe, bump me up a couple hundred kilopascals!" However, awareness of these two measurements is important when making short-term logistical and tactical decisions involving air usage. In other words, making a decision about how to proceed based on the available air supply.

Consider the firefighter that is working in a tight space and trying to make a decision about how to proceed with his air chisel. He calls out to a crew member who is standing near the air cylinder and regulator and asks how much air is remaining in the 4500-psi cylinder. The firefighter on the tool is expecting to hear something like "2000 psi." If the crew member glances down and reads off the SI scale, he would respond with "you got about 14,000." Not only would the information be incorrect, but it can add confusion to an already confusing situation—not to mention aggravating the tool operator.

Low-pressure gauge. Connected on the discharge side of the diaphragm, the low-pressure gauge indicates the pressure being delivered to the hose and tool. Like the high-pressure gauge, it may have two sets of measurement, requiring recognition when setting the pressure.

Relief valve. If the regulator were to become over-pressurized for some reason, the relief valve will open to relieve the excess pressure.

Hose couplings. To connect the hose to the regulator, a hose coupling is located on the discharge side of the regulator. With most tool systems the regulator has a female coupling, but it is possible to see the opposite. The coupling should have some type of safety/locking feature that prevents the coupling from accidentally coming apart if the female coupling is bumped.

Tools, Equipment, and Apparatus

Diaphragm. The diaphragm is the heart of the regulator and is the part that controls the discharge pressure. When the pressure adjustment handle is turned, spring pressure is exerted on the diaphragm that in turn changes the discharge pressure. A clockwise rotation increases the pressure, and a counterclockwise rotation decreases the pressure.

Regulator and bonnet.

Inlet fitting with CGA 347 nut rated to 5500 psi.

O-ring and Allen screw retainer.

If the O-ring is missing, the regulator won't seal properly on the air cylinder.

Threads on a 2216-psi cylinder.

Threads on 2216-psi cylinders are shorter than those on 4500-psi cylinders.

Fig. 2–10 Air tools—regulator components and features

Air hoses

Hoses used for air tools are similar to fire hoses.

- Small diameter hoses supply less volume than large hoses at the same pressure.

- As the length of the hose increases, there is a corresponding decrease in pressure at the tool end of the hose when the tool is operating.

- If the hose is too large or too long, it can become too difficult to handle.

Threads of a 4500-psi cylinder.

Holes in threads of the 4500-psi cylinder prevent a 2216-psi regulator from being pressurized.

Front of regulator.
Max. pressure and type of gas.

Inlet or high-pressure gauge.
PSI is marked on the inner ring.

Back side of the regulator.
High pressure on left, low on right.

High-pressure fittings.
HP = High pressure.

Fig. 2–11 Air tools—regulator components and features

Tools, Equipment, and Apparatus

- Reels can be used to deploy hose quickly.
- Mismatched couplings can create big problems.

Hoses that come with tools should be evaluated before being put in service. During the evaluation process, it should be determined if the hoses are durable enough to use on the crash scene, have an adequate "working pressure" rating, and if they're long enough. If they aren't long enough, consideration should be given to replacing them with better, longer hoses, and keeping the original hoses as back ups.

The T handle adjusts the discharge pressure. Clockwise rotation increases pressure.

The piston regulator has a different look but the same function.

Discharge or low-pressure gauge. psi is indicated on the inner ring.

Locking coupling. To unlock, the ball and groove are aligned.

The locking collar is then pulled back and the hose is connected.

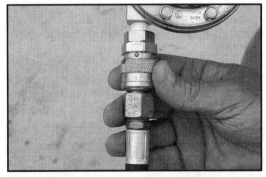

With the hose connected, the collar is rotated to prevent the coupling from unlocking.

Fig. 2–12 Air tools—regulator components and features

When selecting a hose for purchase or specification, the best place to start learning about the options available is at a business that constructs hoses and sells them to commercial users. These companies have the equipment on-site to cut the hoses to length, and attach the fittings and couplings. At the shop where the hoses are made, the counter representative should have a few choices of hose available for examination.

The hose itself should have a tough outer layer to resist cuts on jagged metal and glass, yet be flexible enough to be stored and used easily. The two most important specifications for hoses are the working pressure and the diameter. A working pressure of 300 psi is a good choice because it accommodates most common tools. The 3/8-in. hose diameter is preferred to the 1/4-in. diameter, especially for tools that use a lot of air. Finally, the fittings that go between the hose and the coupling should be commercial grade and attached by machine at the shop. These provide more trouble-free service than fittings that are assembled with a clamp and screwdriver.

Hose couplings

It's best if all extrication air tools and hoses have the same coupling in order to avoid wasting time on the crash scene. A small problem can be encountered with some tools if the manufacturer has used a proprietary coupling on the equipment. Proprietary means the coupling is a special design and is not a common coupling that can be bought in a hose shop. This problem can be overcome by either switching all the air tool couplings to the proprietary coupling or changing the proprietary coupling to a common coupling. Whichever choice is made, the design and specifications of the proprietary coupling should be considered before changing couplings. Consider the diameter of the discharge opening, the volume that the coupling will flow, and any desirable safety features. Additionally, it is a good idea to check with the tool manufacturer to determine if there is any reason the coupling shouldn't be changed to a non-proprietary part.

To avoid problems, it's best to leave tool systems, such as air shores and confined space equipment, intact and with whatever hose, coupling, and regulator set-up is provided by the manufacturer.

Important points for departments converting to 4500-psi SCBAs

As departments move to 4500-psi SCBA systems, those individuals responsible for specifying and using air tools should realize that the change may have an impact on all of their air operated tools. This includes extrication tools, air shores, air bags for rescue, and hazardous materials (HAZMAT), along with confined space equipment.

Older regulators were probably purchased when 2216-psi cylinders were the standard, and may not be designed to be connected to 4500 cylinders. To determine if the regulator can handle the higher pressure, the body of the regulator should be examined as described earlier. The important marking in this case are the words that describe the maximum inlet pressure. If this pressure is less than 4500 psi, the regulator should be replaced. The maximum pressure shown on the inlet side gauge is not a reliable indicator of the maximum allowable pressure.

Tools, Equipment, and Apparatus

Gauges can be swapped out or replaced if broken, making them totally unreliable indicators for maximum pressure ratings. If there are no markings on the regulator, the vendor or manufacturer should be contacted to verify the appropriate pressure rating. Unfortunately, this important discovery may not be made until the equipment is being assembled for a drill, or at a rescue. It's a bad situation to be setting up the equipment at a scene only to find out an outdated, incompatible regulator has been sitting on the apparatus for a month.

If an attempt is made to connect a 2216-psi regulator to a 4500-psi cylinder, the fittings are designed to allow air to leak through a hole drilled in the threads of the 4500 cylinder, giving a clear indication to the user that the components are incompatible. Unfortunately, this safety device will occasionally be overridden by a well-meaning firefighter who switches the nipple and hand wheel on the 2216-psi regulator and replaces it with a 4500-psi nipple and hand wheel. This, of course, is unsafe and is not the proper approach for fixing the problem.

Setting up the system

An air tool system is usually assembled starting with the air source and working back to the tool. For ease of set-up, the air source shouldn't be turned on until the entire system is assembled. By assembling the system first, then charging it, pressure is kept off the couplings—making them easier to connect. As the system is being assembled, check the following key items:

- Is the air cylinder is full? Ensuring that the air cylinder being put into service is full is especially important during extended operations when empty and full cylinders are more prone to get mixed up.

- Are you using all coupling locks? All coupling locks should be used in order to prevent accidental disconnection as the hose is dragged around the scene.

- Have you laid down the air cylinder and regulator? The air cylinder and regulator should be laid down after being connected. If a flat-bottomed cylinder is left upright and is knocked over when under pressure, there is a chance that the regulator or hose coupling may be damaged and start to leak.

- Is the cylinder positioned to provide enough slack hose? The cylinder should be positioned to provide the tool operator with adequate slack hose. If this isn't done and the tool operator runs out of hose during an operation, there will be an irresistible urge to use the hose to drag the cylinder and regulator closer. This shouldn't happen but does. Proper planning and work area set-up is the solution.

Troubleshooting

Problem #11: When the air cylinder is turned on, air leaks at the regulator.

Cause #1: O ring is missing from the end of the nipple.

Solution: Replace the O ring. An Allen wrench may be needed to remove the nipple end. If spare O rings and the correct size wrench are carried in a kit, the regulator can be repaired in the field and immediately returned to service.

Cause #2: A 2216-psi regulator has been placed on a 4500 cylinder.

Solution: Switch to a cylinder compatible with the regulator's maximum inlet pressure rating.

Cause #3: The nipple on the regulator is supposed to be wrench tightened (no O ring) but was hand tightened.

Solution: Tighten with a wrench.

Cause #4: Some air cylinders do not permit some regulator fittings to be tightened as much as they should be.

Solution: Try another air cylinder. This problem is caused by the way the threads on the air cylinder are cut when being manufactured. As the regulator fitting is screwed onto the air cylinder, the fitting will strike the cylinder valve assembly, making an airtight fit very difficult to achieve. If this occurs on the crash scene a little extra effort to create an airtight seal may be possible. It should be determined if the problem is exists with just one cylinder or if the problem is department wide.

Problem #12: When the air cylinder was turned on, the discharge pressure shown on the discharge gauge is higher than desired on the system. Turning the adjustment handle to reduce the pressure doesn't cause the gauge reading to drop.

Cause: The pressure on the discharge side of the regulator diaphragm will remain until some of the air is bled off.

Solution: Turn the regulator adjustment handle in the direction used to lower the pressure, and then briefly operate the tool to bleed the air off the system. Once the regulator is bled off, it can be readjusted to the desired pressure.

Problem #13: With the regulator set at the correct operating pressure, the tool operates correctly for a couple of seconds then the speed or power of the tool drops off.

Cause: Inadequate volume or pressure being delivered to the regulator.

Solution #1: Open the air cylinder fully to allow a higher volume of air to flow.

Solution #2: Check to see if the air cylinder is in the 0–800 psi range. While a tool may only require 95 psi to operate, the low cylinder pressure will cause the air delivery rate to drop off, resulting in a slower tool speed. If the cylinder is low, preparations should be made to switch to a fresh cylinder.

Air chisels

Air chisels were one of the first tools used for extrication but fell out of favor with the introduction of the heavy hydraulic tools. Fortunately, their value has been rediscovered, and once again, they are considered by many to be one of the primary tools for extrication. While air chisels lack the brute power of the hydraulic tools, they are valuable when it's necessary to quickly remove metal in an operation. Removal of metal, one of the four primary tactical options for extrication, can provide the crew with the necessary space to perform victim access and removal. While hydraulic shears are very useful for cutting roof posts and making relief cuts, they are of less use when long, continuous cuts are required. When performing evolutions like the third-door, the ability to make long, curved cuts make the air chisel the tool of choice. When combined with other tools, the air chisel can play a valuable role in reducing the time required to extricate the victim.

How an air chisel works. The easiest way to understand how an air chisel works is to think about how an ordinary chisel works. An ordinary chisel cuts wood or metal when the chisel is struck by a hammer that forces the sharp edge of the chisel through the material. An air chisel does the same thing, but instead of swinging a hammer, air pushes a piston through a cylinder that in turn strikes a chisel bit, but at a much faster rate than a hammer can be swung.

When the back end of the chisel bit is struck by the piston, the cutting edge of the chisel is driven through the material. The frequency at which the piston strikes the bit is measured in blows per minute (bpm). Specifications for air chisels often include the number of blows per minute at a specified operating pressure. After the piston strikes the chisel bit, the chisel bit is pushed along until it is stopped by the bit retainer.

The retainer is the device that allows the bits to be changed quickly and prevents the bit from flying out of the tool when the trigger is depressed. There are a few types of retainers, but the one that is most useful is the quick change, or retractable collar bit retainer. To insert or remove the chisel bits, the collar on the retainer is pulled back and the bit is removed or inserted. Other types of retainers will work, but they're not as quick or easy to use.

Before picking up the air chisel and starting to cut, the tool operator should size up the material so the proper bit can be inserted into the gun. There are several choices when it comes to bits, so it's important to understand the function of each bit.

To understand the function of each bit, it is helpful to categorize the types of materials that may need to be cut.

- Smooth, single-thickness sheet metal
- Crumpled, single-thickness sheet metal
- Double/triple-thickness sheet metal
- Door hinges or latch bolts

To simplify this list even further, the last three categories can be grouped together, resulting in a total of two broad categories: smooth, single-thickness sheet metal, and crumpled (or thick) metal.

T and dual cutter bits. Smooth, single-thickness metal can be cut with a T cutter, or a dual cutter. These are great bits for smooth metal because both bits have built-in guides that help the cutting portion of the bit maintain its proper position and depth while cutting. The vertical part of the bit, the part that does the cutting, is actually blunt. This blunt surface will easily cut through the thin metal, but when it hits two or more layers of metal, the speed of the cut is reduced dramatically to the point of being ineffective. Even with its limitations, the bit is a good choice when a lot of smooth metal needs to be cut, as in the case of entry though the roof of a car resting on its side. The T cutter is also good choice for those individuals using the tool for the first time. The guide on the T cutter is larger than the dual cutter, making it easier to control.

Curved cutter bit. If the metal isn't thin and smooth, the curved cutter is the bit of choice. With its wide, sharp edge, the curved cutter is capable of cutting through multiple layers of sheet metal, door hinges, and latches. The drawback to the curved cutter is that it requires some skill to be used effectively. With experience, however, the curved cutter can be used in all situations—almost eliminating the need for the other style bits. This bit provides a direct correlation between practice and effectiveness. Without practice, the tool operator will have a very difficult time controlling the direction and depth of the cut.

Ordinary or cold chisel bit. If bolt heads or nuts need to be cut off, the ordinary bit that resembles a cold chisel can be used. The edge of the ordinary chisel bit is thicker than the curved cutter, giving it greater strength and durability. The cutting edge is beveled on both sides like the curved bit, but unlike the curved bit, the cutting edge is straight. This straight edge provides for better positioning of the center of the blade when cutting off bolt heads. This is an advantage over the curved blade because the concave shape of the blade often prevents the center of the blade from making contact with the bolt.

Making the basic cuts with the T cutter. When starting out with an air chisel, performing intricate, complicated cuts isn't as useful as first learning to make straight easy cuts with good control. The simple straight cuts provide the operator with the feel for the tool that will be needed later for precision cuts. To get a feel for the tool, the following steps will help:

1. Assemble the system, making sure the air cylinder is opened fully. If not opened fully, the volume delivered to the gun will be decreased, which in turn reduces the speed and power.

2. Insert a T cutter in the gun. If a T cutter isn't available, a dual cutter can be used.

3. Select a practice area on the vehicle that has a single thickness of sheet metal. Good choices include fenders, doors, and roofs.

4. Place the cutting edge of the bit against the metal and lift the body of the tool away from the metal so when the trigger is pulled the angle won't cause the bit to slide along the surface.

5. Pull the trigger briefly, just long enough that the cutting edge of the bit penetrates the surface of the metal.

6. Lower the gun so that it is close to the metal surface. This brings the built-in guide of the T bit into proper alignment with the metal.

7. Rotate the body of the gun so that with the bit in position, the hose is positioned off to the side. The idea is to keep the hose out of the sharp metal and to get the handle of the gun out of the way. For best control, if the tool operator is right handed, the gun should be positioned so the hose is to the operator's right, not left. By positioning the hose to the right, the operator's hand is maintained in a more natural position, providing greater range of motion and control.

8. Squeeze the trigger while exerting moderate forward pressure on the gun. If the bit doesn't move forward, usually one of the following things is occurring:

   - The air cylinder isn't turned on all the way.

   - The pressure isn't set properly. (A good starting pressure is about 95-psi.)

   - The area is reinforced with more than one layer of sheet metal

   - The operator isn't exerting enough pressure on the gun to push it forward through the metal. This is usually the problem. After a while an experienced tool operator will be able to hear the distinctive double tapping sound made when there isn't enough force

Tools, Equipment, and Apparatus

being used on the gun. The double tapping sound occurs when the gun is pushed backward instead of the bit being pushed forward.

9. Once the cut is started, the operator should attempt to make a smooth straight line. A couple of these straight cuts will give the operator a basic feel for the gun.

10. The operator should then try to make some smooth curves.

11. Feel for the cuts. A marker can be used to draw ovals or circles on the practice vehicle. At this point, the operator is concentrating on control. This is an important skill, especially when the air chisel is used in tight areas and close to a victim. It would be easy to say that the air chisel shouldn't be used close to a victim, but that's simply not reality. In fact, in many cases, when working close to the victim, the air chisel may be the only tool suitable for the situation.

Making cuts with the curved cutter. Once the tool operator has developed the feel of the air chisel and the T cutter, it's time to try the curved cutter. While it's more difficult to use when first learning, the curved cutter is the bit of choice for experienced firefighters. The big advantage of the curved cutter is that it will cut everything a T or dual cutter can cut, and more. This means that the experienced crew member using the air chisel may be able to use one bit for an entire extrication, eliminating the need for switching bits. The downside to the curved cutter, in addition to the skill required, is that the bit can be dulled, requiring additional maintenance, or replacement during long operations.

Following are steps to help get started with the curved chisel bit:

1. Select a curved cutter bit and insert into the tool. If given a choice of different lengths choose the longest bit. Usually the longest bit is about 18 in. long.

2. Place the corner of the bit against the metal while the entire tool is positioned so the bit won't skip along the surface when the trigger is depressed.

3. Grip the bit just behind the flattened part of the bit to help maintain good control. At first, this close positioning of the hand may prompt questions about safety and the risk of being cut. Experience has shown that close positioning of the hand has several advantages.

    - The cutting edge of the bit can be better controlled with a hand located closer to the cutting edge. Better control means safer operation.

    - If the operator's hand isn't positioned close to the end of the bit, it will probably be positioned over the bit retainer. This positioning leads to accidental movement of the retainer collar, which in turn allows the chisel bit to be fired out of the gun like a spear.

    - With proper PPE, the chance of being cut is minimized.

4. Depress the trigger (with the corner of the bit against the metal), until the bit makes a cut about 3/4 in. long.

5. Rotate and position the entire tool so the flat part of the blade is lying almost flat

against the surface of the metal to be cut. If the blade won't lie down flat, position it as close as possible to this position.

6. For good control, more than half of the cutting edge should be visible and outside the cut. If more than half of the cutting edge is allowed to slide into the cut, there is a greater chance of the whole blade sliding into the cut and getting buried.

7. Position the gun so that the barrel is close to the surface of the metal. Cut with the hose out to your dominant side. For exam-

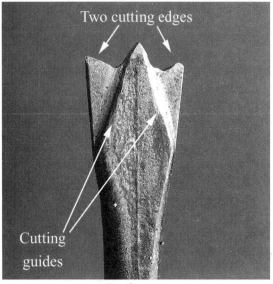

Dual cutter

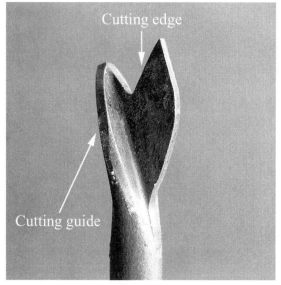

T cutter

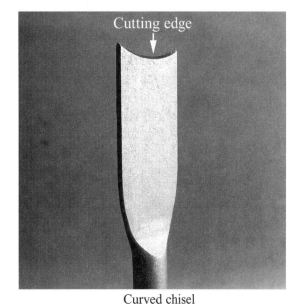

Curved chisel

Cold chisel

Fig. 2–13 Air chisel bits

Tools, Equipment, and Apparatus

ple, if you are right handed, the hose should be positioned off to your right side, not left.

8. Start the cut while observing and concentrating on how and why the cut changes direction as the position of the gun is changed.

9. Establish a basic feel for the gun. Attempt cutting a straight line.

Like the curved cutter, the T cutter requires a steep approach when starting the cut.

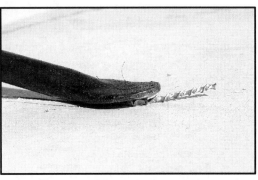
After the initial cut is made, the bit is lowered so the guide runs along the surface of the metal.

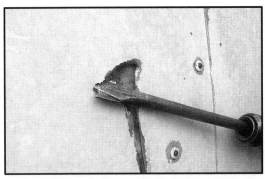
The dual cutter has two cutting edges, and is used in the same manner as the T cutter.

The dual and the T are useful when smooth single thickness sheet metal needs to be cut.

The cold chisel bit can get under screw and bolt heads.

The wedge shape pushes the screw head up then cuts the screw shaft and threads.

Fig. 2–14 Air chisel basic cuts—the T, dual, and cold chisel bits

Following are typical problems that can occur when cutting with a curved bit:

- Instead of cutting straight, the cut develops a curve or angles off to the side. These problems occur when operator doesn't have the bit and the gun lined up properly. The problem can be fixed by repositioning the gun until the bit cuts in a straight line. To help understand the subtle movements needed for accurate straight cutting, some test cuts can be made. When making the test cuts, the operator should make some exaggerated changes in the position of the gun to see how the changes affect the cut.

- When making a cut, the blade slips off the metal and ends up buried. This problem occurs when the barrel of the gun is moved away from the surface of the metal being cut. As the angle between the surface and the tool increases, the bit is more likely to skip off the cut.

- When trying to make a turn or curve, the bit continues in basically a straight line. This problem can be corrected by swinging the gun in a wider arc along the surface of the metal being cut. A good comparison can be made between cutting a curve with an air chisel and changing the direction of an attack line. The movement is made a lot easier when the hose behind the nozzle is moved around to get the nozzle lined up on the target.

Lubrication. The tolerances between the piston and the cylinder of an air chisel are tight, requiring lubrication for proper performance. Without lubrication, wear on the internal parts will increase, or the piston may stop moving entirely. Lubrication is provided by the applying of a few drops of pneumatic tool oil down the barrel with the bit removed. Some manufacturers may recommend oiling through the intake side of the tool, others won't. If there is any question, the manufacturer of the tool should be contacted for guidance. Generally, lubricating down the barrel is adequate because the oil is applied directly to the piston and cylinder. After adding a few drops of oil down the barrel, the gun can be shaken to help distribute the oil between the piston and the cylinder. As the gun is shaken, the piston can be heard moving up and down inside the cylinder indicating adequate lubrication. It's a good idea to lubricate the tool before and after every training session, and after every use at a crash scene.

Lubrication before use at the crash scene isn't necessary if normal maintenance of the tool has been performed. To promote the proper maintenance of the equipment, a small plastic squeeze bottle of oil can be stored on the apparatus with the tool. A one-ounce container holds enough oil to last a couple of years under typical conditions and use.

Maintaining the chisel bits. The T and dual cutter bits require no maintenance other than wiping down with any oily rag to prevent rust. Curved cutters and cold chisel bits should be sharpened after every use, whether used in training or at the emergency scene. The difference between a sharp, curved bit and a dull one is significant and can make or break an operation.

Learning to sharpen a curved bit only takes a few minutes, and can result in an edge comparable to the one ground at the factory. All that is required is a grinder with a 5- or 6-in. grinding wheel, a cup of water, and an oily rag. The following steps will provide a sharp edge quickly:

- Ensure that the grinder has a smooth, uniform surface across the face of the wheel.

Tools, Equipment, and Apparatus

If the wheel is damaged or has grooves in it, the wheel should be replaced or repaired with a wheel dresser. A wheel dresser is an inexpensive hand-held tool that restores a smooth surface to the wheel.

- Adjust the tool rest so that there is a very small gap between the rest and the wheel. By maintaining a small gap between the rest and the wheel, the chance of the bit getting wedged between the wheel and the rest is reduced.

For best results, the hose is positioned off to the operator's side.

If not positioned to the side, the grip or hose may strike the object being cut.

To start the cut, the bit is positioned at a steep angle.

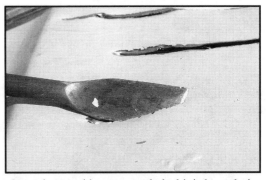

Once the metal is penetrated, the bit is brought in closer to the object being cut.

The bit will rotate in the cut and lay almost flat against the surface. This provides good control.

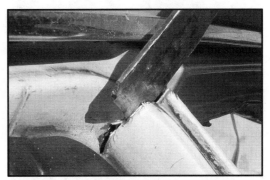

Whenever possible, at least half of the bit should remain outside the cut.

Fig. 2–15 Air chisel basic cuts—the curved cutter

Vehicle Extrication: A Practical Guide

Tool rest is set close to wheel.

Clear, clean shields help prevent eye injuries. Safety glasses should also be worn.

The shaft of the bit is positioned in a horizontal position.

This shaft is positioned above horizontal.

This shaft is positoned below horizontal.

The next angle to set will create the bevel The best angle is about 30 degrees from the wheel.

Fig. 2–16 Sharpening air chisel bits—curved cutter

- Adjust the clear shield properly and don eye protection. Clothing and gloves that may pose an entanglement hazard should not be worn when sharpening. Bulky and loose articles of clothing have a greater chance of getting caught in the machine.

- Turn on the grinder and bring it up to operating speed.

- Grasp the shaft of the bit with the hand furthest away from the grinder. The closer hand is used to help support and control the bit near the flat portion of the blade.

Tools, Equipment, and Apparatus

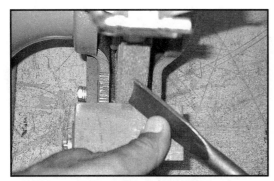
This angle is about right.

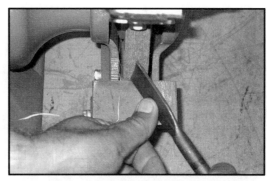
This angle is too large and will not create an edge that is sharp enough.

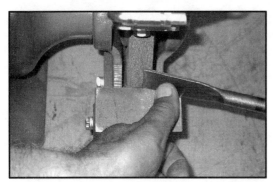
This angle is too small resulting in a edge like a knife that will dull quickly.

When set correctly, the entire edge of the bit will contact the wheel.

During sharpening, the bit should be cooled with water to prevent overheating.

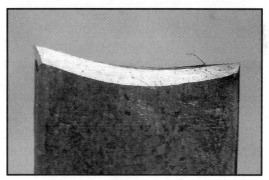
The right edge of this bit was overheated, indicated by discoloration.

Fig. 2–17 Sharpening air chisel bits—curved cutter

- Position the bit so that the shaft is horizontal—not tilted up or down.
- Place the shaft in the horizontal position, and set the angle of the bevel. The idea is to create a bevel with the proper sharpness while, at the same time, maintaining the curve of the cutting edge. The easiest way to do this is to bring the bit close to the face of the wheel, while keeping the shaft parallel to the face of the wheel. With the edge of the bit about 1/2 in. away from the wheel and maintaining that distance, the shaft is pivoted about 30° away from the wheel. With the edge of the bit

Vehicle Extrication: A Practical Guide

This edge is uneven because it was held at different angles while being sharpened.

This edge is fairly uniform in shape and is acceptable for sharpening done by hand.

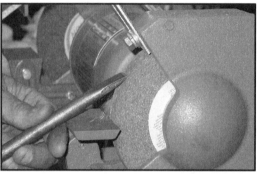

When the cold chisel bit is sharpened, the bit is in front of the grinder instead of the side.

The angle of the bevel is set by the position of the bit on the wheel.

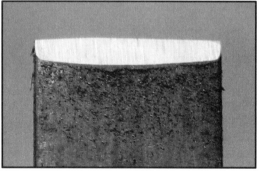

Unlike the curved cutter, the edge of the cold chisel is straight and uniform.

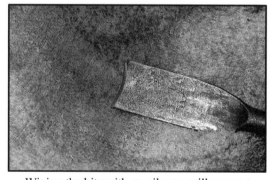

Wiping the bits with an oily rag will prevent rust from developing.

Fig. 2–18 Sharpening air chisel bits—curved cutter and cold chisel

about 1/2 in. away from the wheel and maintaining that distance, the shaft is pivoted about 30° away from the wheel.

- Move the whole bit toward the wheel while maintaining the proper angles. The existing bevel and curve of the bit should come close to matching the one that will be set by the wheel when it makes contact. The bit is gently brought against the wheel.

Tools, Equipment, and Apparatus

- Make contact with the wheel. If the angles look correct, move the bit from one side of the wheel to the other. This movement eliminates the creation of a groove in the wheel.

- Make a few passes and dip the bit into water for a few seconds to keep it cool. Overheating the blade will cause the metal at the edge to turn blue, purple, violet or brown indicating that the bit has been overheated. Overheating can cause the bit to become soft and easily dulled. In between passes, it's best to keep the cutting edge of the bit at a temperature that allows it to be handled by hand without burning the user.

- Repeat passes as needed to bring the edge to a uniform appearance. As the passes are made, the bit will need repeated cooling to avoid being damaged.

- Examine the blade. The bevel of the edge should be sharp enough that it will cut through metal but not so sharp that it will bend over when used to cut thick metal. The best comparison that can be drawn is that of an ax that that has just been sharpened.

- Observe the edge. If it looks good, dry and wipe down the bit with an oily rag. The oil will prevent the freshly sharpened edge from rusting.

Sharpening a cold chisel bit can be easily accomplished by resting the bit against the front of the tool rest and bringing the edge of the bit close to the wheel. If the angle created looks like the angle that came from the factory, the bit is lowered to the wheel, and the edge sharpened. This edge should also be wiped down with an oily rag to prevent rust.

Impact wrenches

An impact wrench may be needed on occasion when a heavy bolt needs to be removed. Impact wrenches are simple to understand and have limited adjustment. There will be some type of switch that controls the direction of the tool—either forward or reverse. Like the air chisel, the impact wrench requires lubrication, and lubrication points may be marked on the tool. When specifying and purchasing sockets for an impact wrench, only sockets designed for use with impact wrenches should be used. Standard hand tool sockets won't hold up to the pounding action of an impact wrench and shouldn't be used for both practical and safety reasons.

Whizzer/die grinder

The whizzer is a useful tool that functions like miniature ventilation saw that's been outfitted with an abrasive wheel. It's useful for cutting metal in tight quarters where other more powerful tools won't fit. Just like other saws, the use of an abrasive blade on ferrous metal will cause a shower of sparks. If it's necessary to use this tool during extrication, steps should be taken to minimize the risk of fire. This could include removing combustibles from the work area, wetting down the area before starting to cut, and having adequate fire suppression equipment in place in the event of a fire. If the tool is used and a fire does start, fire suppression activities should be very aggressive to stop the progression of the fire while the victim is trapped.

Operating the whizzer. The design of the tool is simple enough, but operating it efficiently requires attention to a couple of details.

- Most whizzers are equipped with a simple trigger and safety combination lever. The safety needs to be disengaged to activate the trigger. This is done by simply sliding the safety lever to disengage it and then depressing the trigger lever. Once the safety is released and the trigger lever partially depressed, the safety will remain disengaged. When pressure on the trigger lever is released, the safety is repositioned.

The whizzer is useful in spaces that are too tight for other tools.

A drawback to the whizzer is the amount of sparks produced when cutting steel.

When cutting aluminum, sparks aren't a problem.

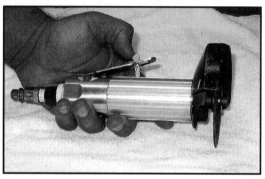

To operate the whizzer, the safety is released by pushing the lever forward.

As the safety is released, the operating lever can be depressed.

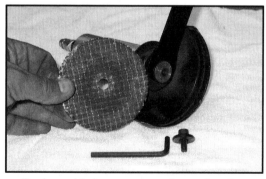

To change blades, an open-end wrench and an Allen wrench are needed.

Fig. 2–19 The whizzer

This type of safety is fairly low tech and there may be an urge to defeat or remove the safety lever to make the operation of the trigger easier. Removing the safety will cause an unnecessary safety problem any time the tool is set down or bumped. As is the case with most safety devices, defeating the device can cause serious problems and injuries.

- To avoid binding the blade in the kerf (the narrow gap created when the metal is cut), the tool operator may want to brace an arm against a solid object.

- Like a ventilation saw, a whizzer makes a distinctive sound when it starts to bind or slow down. The operator should be attentive to the sound of the tool, and either exert less pressure on the tool, or reposition the saw to eliminate any binding.

Changing and attaching blades to the whizzer. Use the following guidelines when attaching blades:

1. Disconnect the tool from the air source to prevent accidental activation of the tool during the blade installation.

2. Attach or change the blade by inserting a wrench behind the blade to prevent the shaft from spinning.

3. Position the wrench and loosen the screw at the end of the shaft with an Allen wrench.

4. Remove the screw and washer.

5. Install the blade, lining up the shaft with the hole in the blade.

6. Reinstall and tighten the screw and washer.

7. Spin the blade by hand to ensure that it spins true and doesn't wobble.

8. Attach the tool to the air source.

Like the other pneumatic tools, periodic lubrication with pneumatic oil is necessary to keep the tool running at maximum efficiency and to extend the life of the tool.

Air bags

Air bags may be used in the extrication setting when it's necessary to lift a vehicle to free a victim. The procedures used in this type of environment are the same as other settings, but the following tips can help with making an effective lift:

- Avoid sharp edges and exhaust systems to reduce the chance of damaging the bags.

- Use some type of bag protection if sharp edges or the exhaust system can't be avoided. Mud flaps and sections of old conveyor belts provide good, durable bag protection. In the worst-case scenario, when the lift must be made over some sharp objects and there is no edge protection on the scene, it may be necessary to sacrifice an unused air bag and use it as edge protection. This bag shouldn't be inflated and may have to be replaced after the incident is concluded. A risk/benefit analysis will usually show that the willful destruction of department property is justified when a life hangs in the balance.

- Stack two bags, even when it looks like only one bag is needed to gain the height

Vehicle Extrication: A Practical Guide

needed. If there are plenty of bags available, there's no harm in having a second bag available if a little extra height is needed for the lift.

- Determine if there are going to be two lift points. Four bags will be needed.

When it appears only one air bag will be needed, two should be stacked if available.

When four bags are needed, four different color hoses should be used for clarity.

The operator should be in a position to see and hear the officer and crewmembers.

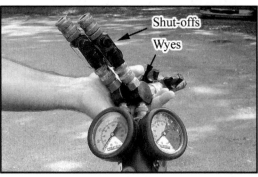

To operate four air bags, two wyes and four shut-offs are attached to two discharges.

When using four bags, each gauge will be used for two bags.

Each set of inflate/deflate buttons will also control two air bags.

Fig. 2–20 Air bag lifting operations—the four bag set-up

Tools, Equipment, and Apparatus

If a four-bag system is to be built, along with the four air bags, four hoses, shut-offs, and discharges will be needed. Most air bag controllers are designed with two sets of controls and gauges allowing two air bags to be lifted simultaneously or independently of one another. Some controller kits contain a dual controller along with two wyes, four shut-off/relief valve attachments, and four hoses. To build a system that provides independent control of four bags with a two-bag controller, take the following steps:

1. Insert a wye into each of the two discharge couplings on the controller.

2. Attach the four shut-offs to each of the wye discharges.

3. Attach a different color hose to each of the four shut-offs.

4. Operate each bag independently. Close all of the shut-offs—except the one attached to the bag that is to be inflated. To inflate the bag, depress the inflate button on the corresponding side of the controller.

5. Inflate two bags simultaneously. Open the two appropriate shut-offs and depress the corresponding inflate buttons.

6. Avoid confusion and unintentional inflation by closing the shut-offs after each inflation procedure.

7. Set up the system so that each bag is controlled by its own side of the controller if the plan calls for two of the four bags to be inflated simultaneously. This avoids the problem of having to inflate both bags with one control button, while trying to feather the lift with the two shut-offs.

If this seems a little confusing it may be easier to view the way two gated wyes would be used on two fire hoses. There would be two supplies to the wyes, four hoses coming off the wyes, and four valves to control the flow to each of the four lines.

Electric Tools

Until recently most electric power tools in the fire service were supplied by generators mounted on an apparatus. Recent improvements in battery-powered tools have expanded the role of electric tools, especially those with converters that allow the tool to be used with batteries or AC power. Battery-powered tools have a limited operating time, making the option of converting to AC power a plus.

The reciprocating saw has been recognized as an important extrication tool. These are fairly simple tools with few adjustments. If there has been one complaint in the past about reciprocating saws, it was the durability of the blades. Fortunately, the size, design, and durability of the blades have improved recently, elevating the reciprocating saw into the class of primary extrication tools.

When selecting blades, there are usually questions about which tooth pattern and count should be used. When cutting metal, most shop texts recommend using blades with at least 14 teeth per inch (TPI). This recommendation is made in the interest of extending the life of the blade, and for the cleanness of the cut. Through experience, most firefighters have determined that cutting with such a blade is a slow process. Most experienced firefighters now prefer what is called a *more aggressive blade*—one that has 10–14 TPI. The justification for this choice is that blades are cheap and that rough cuts have no negative impact on a trapped victim.

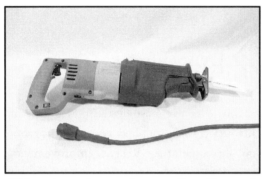

This reciprocating saw features adjustable speed control and trigger and an adjustable shoe.

The stroke per minute speed is controlled by this wheel. 1=slow, 5=fast.

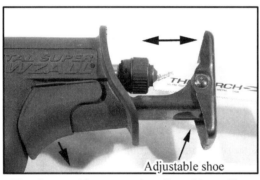

The shoe on this saw can be adjusted by pulling down on the shoe release lever.

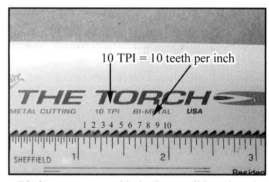

Blades are measured in both overall length and the number of teeth in 1".

An electric impact wrench is an alternative to the pneumatic impact wrench.

Battery-powered tools can provide an option when engine-powered equipment isn't feasible.

Fig. 2–21 Electric tools—reciprocating saws, impact wrenches and spreaders

Tools, Equipment, and Apparatus

Manual Hydraulic Tools

The most commonly used manually operated hydraulic tools include the following:

- Jacks
- Manually operated spreaders
- Manually operated cutters

All manually operated hydraulic tools receive their power when the operator applies force to a handle or foot pedal, which in turn operates a piston in a cylinder that increases hydraulic pressure. This force is then transmitted to the parts of the tool that are doing the work. When the operator exerts enough force on the handle or pedal to meet the maximum operating hydraulic pressure of the tool, a relief valve operates to prevent over pressurization.

Manually operated hydraulic tools consist of one or two components. In the case of one-piece units, the pump, fluid reservoir, and the operating part of the tool are all in one piece. Two-piece tools have a pump/reservoir, the operating part of the tool, and a hose that connects them. When using a two-piece tool, the pump/ reservoir component should be kept flat and level when being operated. This is necessary because some pumps only operate properly if kept flat. This can be a problem in situations when the tool is being operated in an area that requires the pump to be held up off the ground by the pump operator.

When servicing hydraulic tools, it is very important that the correct hydraulic fluid is used in the tool. Some tools have labels identifying the correct fluid, while others include the information in the owner's manuals. If there is a question about the correct fluid to use and there is no manual available, view the manufacturer's web site, or contact their representative for the information. Once the correct fluid is determined, use paint pens, engravers, or heavy-duty tags on the tool to indicate the correct fluid to use. The next crew members who aren't sure which fluid to use will be thankful for the effort. This small step can eliminate costly repairs—or worse, tool failure at the crash scene.

Bottle jacks

Bottle jacks represent one of the simplest forms of manually operated hydraulic tools. The typical, simple, bottle jack consists of only a few parts.

- Base—designed to give the tool a wider stance to keep the tool in the vertical position.
- Piston—the part that moves up and down in the cylinder to create pressure.
- Cylinder—the empty tube the piston moves in.
- Valve—controls the direction and movement of the hydraulic fluid.
- Handle—gives the operator some leverage (mechanical advantage) when pumping the fluid.

Basic jack operation— to raise

- Move the jack into position.
- Move the valve into the closed position. Usually this is accomplished by a clockwise rotation of the valve or a handle that operates the valve.

- Place the handle into the handle receiver and operate downward to create the pressure that forces the piston upward.

- Observe the handles of manually operated jacks. Some may have two notches at one end of the handle that can be used for closing the valve.

- Rotate the handles on simple floor jacks clockwise to close the valve.

Basic jack operation—to lower

- Rotate the valve counter clockwise to open the valve.

Floor jack

Some heavy-rescue units carry floor jacks for lifting operations because of their stability, power, ease of positioning, and operation. They have a low profile, which allows them to slide beneath objects that have limited vertical clearance. The wheels on a floor jack allow it to be moved easily on hard surfaces, but are difficult to move and position on soft surfaces. The tool is rather long, which allows the jack to slip under an object, but still allows the handle to have full range of motion. The lifting pad on the jack usually has some type of shape/surface that helps grip the object that is being lifted. Additionally, the pad can be rotated (if necessary) for fine adjustments between the contact points of the tool and the object.

Basic floor jack operations—to raise

- Line up the jack so that the handle can be used to push the jack forward, beneath the object to be lifted. Floor jacks are difficult to move sideways, making the approach to the lifting point important. If the positioning turns out to be off quite a bit, it's best to pull the jack out, line it up for a proper approach, and then slide it under the object again.

- Position the jack and rotate the handle clockwise to close the valve.

- Raise the floor jack by moving the handle up and down, which operates the pump.

Basic floor jack operations—to lower

To lower the floor jack, turn the handle counterclockwise, which opens the valve. Most jacks allow the handle to be rotated and the valve to be operated with the handle in any position.

Even though the floor jack appears to be a large simple tool, good floor jacks have a significant amount of fine control when raising and lowering. Raising the jack in small increments is directly controlled by the downward motion on the handle, which is easy to regulate. The handles on floor jacks are long, giving the operator the ability to move the handle down a fraction of an inch, which results in a very small amount of lift.

Lowering the jack isn't quite as precise as lifting because there is less leverage involved in opening in the valve, which means less control. To lower the jack with good control, the operator

Tools, Equipment, and Apparatus

should use two hands on the handle, get a good grip, and slowly rotate the handle counterclockwise in the smallest increments possible. The two-handed grip is necessary because the valve is likely to go from the firmly closed position to the open position very quickly. By having a good grip, the operator can resist the counterclockwise rotation of the right hand with the left hand.

One way to reduce the chance of a sudden movement to the open position is to use the proper amount of force to close the valve at the beginning of the operation. The valve should be closed firmly enough to close the valve, but not so firmly that it will open suddenly when the handle is rotated counter clockwise. If too much force is used when closing the valve, it may be difficult to open under control; if too little force is used to close the valve, the jack may momentarily extend, only to settle down again as the fluid leaks by the valve. Practice and training will provide control and a good feel for the jack.

Manually operated spreaders

There are several variations of the manually operated spreader tool. The earliest variations have been borrowed from the automobile body repair industry. These early spreader tools were used in the period of history after crowbars and wreckers were considered the primary tools for extrication and before the first engine-driven hydraulic tools were put into widespread use. More powerful adaptations of the early manually operated spreader tools are now available and perform well in simple extrication settings.

Automotive repair-style spreaders

One of the earliest adaptations of these spreader tools actually consisted of two spreader tools carried in one kit. With this set up, the idea was to create a purchase point with the smaller of the two tools, then insert the larger tool to complete the procedure. This tool system required two operators—one to position the spreaders and one to operate the hand pumps. Coordination between the two operators was the responsibility of the crew member who was positioning the spreaders. The spreaders and hand pumps were different colors—a convenience that allowed the crew member positioning the spreaders to call out, "Down on the red, up on the yellow," or "Down on the yellow, up on the red."

When the pump operator received the command to go down on the yellow, the valve on the yellow hand pump would be opened, allowing the spring in the spreader tool to close the spreader arms. When the command was given to go up on the red, the valve on the red hand pump would be closed and the operator would start pumping the hand pump. This procedure was continued until the objective was achieved or until it was determined that the spreaders didn't have the necessary power to complete the task.

Modern manually operated spreader tools

Recognizing that all fire departments couldn't afford engine-driven hydraulic tools and that many, if not most, door-popping operations could be completed with manual hydraulic tools, manufacturers started improving the design of manually operated hydraulic spreaders. Improvements made to these tools resulted in the following benefits:

- The ability for most of these spreaders to be operated by a single crew member.
- Increased power over their predecessors.
- Increased resistance to leaks.
- Better tip design.
- Better valve control design.
- Relief valves to prevent over-pressurization.

Basic spreader operations—to spread

- Move the valve handle to the "spread" position, as indicated by the label near the valve.
- Operate the pump handle.

Basic spreader operations—to close

- Move the valve handle to the "close spreader" position, as indicated by the label near the valve.
- Operate pump handle, if necessary.

Manually operated cutters

Manually operated cutters were developed to enhance the system created by the manually operated spreaders. The operation of cutters is similar to the spreaders.

Basic operation— to cut/close

- Move the valve to the "cut/close" position.
- Position the object to be cut as deep as possible in the cutter blade to maximize the cutting force.
- Operate the handle to close the cutter blades.

Basic operation—to open

- Move the valve to the "open" position.
- Operate the handle to open the cutters.

ENGINE-DRIVEN HYDRAULIC TOOLS— A HISTORICAL PERSPECTIVE

During the 1970s, engine-driven hydraulic tools moved into the limelight with their powerful spreaders and engine-driven pumps (see Fig. 2–22). These engine or motor-driven tool systems are also commonly referred to as heavy hydraulic tool systems or were generically referred to as the Jaws of Life—a name used to identify the Hurst brand of rescue tools. The early spreader tools were heavy and fatiguing to use, but they were far superior to the manually operated hydraulic tools that preceded them. The typical spreader tool weighed over seventy pounds, but was fairly well balanced and easy to use. In comparison to today's tools, they seem heavy, but at

Tools, Equipment, and Apparatus

the time they were considered to be (and truly were) marvels. This period in the history of hydraulic rescue tools was a real turning point in terms of equipment and techniques used in vehicle extrication.

Probably the most common tool set at the time was built by Hurst and consisted of a pump powered by a two-stroke engine, a length of hydraulic hose that was connected to a power plant, and a 32-in. spreader tool. The most common pumps developed a fluid pressure of 5000 psi. To enhance the capabilities of the spreaders, cutter attachments were available at the time that attached to the spreader arms. These cutters have been replaced by specially designed cutting tools.

Typical two-stroke powerplant that was in widespread use in the 1970s and 1980s.

These 32-inch spreaders were often paired up with the two stroke powerplant.

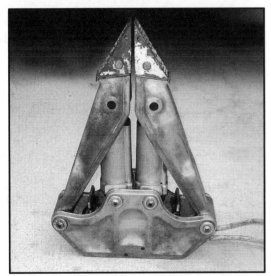
Early dual piston spreader tool by Hurst.

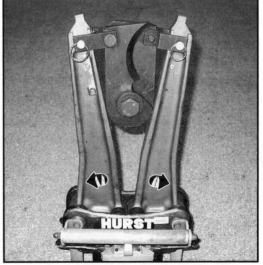
Early cutter attachments were replaced by separate cutter tools.

Fig. 2–22 The early hydraulic rescue tool system

When the new cutters came on the market, crew members working at crash scenes started having problems with the hose and switching tools. The most common problem was when it was necessary to switch tools at the end of the hose that was plumbed into the power plant without the use of a dump valve. To switch tools, it was necessary to turn off the engine on the power plant before the hoses were disconnected, disconnect and connect the tools, then restart the power plant. If the engine wasn't shut down prior to disconnecting the hose, as was often the case, the hose connected to the discharge side of the pump would become pressurized to 5000 psi. When attempts were made to connect the female coupling of the tool to the male fitting on the hose, the pressure behind the check valve made it nearly impossible to connect the couplings.

If this occurred during training, a wrench would be used to back the threaded coupling off of the discharge fitting on the pump to relieve the pressure in the hose. The objective was to provide a very small opening so the pressure could be relieved as the fluid came out the opening. While this was a reasonable solution on the drill ground, at the crash scene a quicker method was commonly used when the system accidentally became pressurized. A crew member in full protective gear would wrap a rag around the male hose fitting and then slam the fitting against a firm, hard surface. The pressure would be released immediately and the rag was supposed to capture the majority of the fluid before it sprayed into the air and became vaporized. This seemed like a good idea at the time, but more than one firefighter has received a face full of hydraulic fluid using this method.

Fortunately, the introduction of the dump valve solved most of these problems. The dump valve reroutes fluid from the discharge side of the pump back into the reservoir—instead of to the hose, which prevents the hose from becoming pressurized. This was considered a major improvement and led to the retrofitting of many pumps that were in service. The dump valve, or selector valve, soon became a standard feature on all manufacturers' power plants.

The next tool to become popular in the fire service after the spreaders and cutters were the rams. The fact that they were introduced after the spreaders and cutters is ironic because the rams are the simplest of hydraulic tools consisting of a piston, cylinder, and control valve. A ram is just a large version of the parts that make a spreader and cutter operate. The most likely reason for their late introduction is that nobody had recognized a need for them. The dash roll-up procedure, which uses rams to push the dashboard forward, is probably an adaptation of the earlier versions of the dash lift and dash roll-up procedures that used the spreaders as a pushing device.

Other devices became available that enhance the performance and set up of the heavy hydraulic systems. These accessories include the following:

- Chain and shackle kits
- Accessory kits for spreaders and rams
- Hose reels
- Manifold blocks

The chain and shackle kit was one of the first accessories to be used with the spreaders, and later with rams. The chain kit enables the spreaders to be used as a pulling tool in addition

Tools, Equipment, and Apparatus

to being used as a spreading tool. In the 1970s, the chains were considered critical in the extrication of victims trapped behind the dashboard and steering wheel. The steering wheel could be pulled away from the occupant by anchoring one chain to the chassis in the front of a car, and the other chain around the steering column, and then closing the spreader arms. While this evolution is no longer the first choice in severe frontal crashes, it has certainly saved a lot of lives.

The chain-and-shackle kits have also been used with rams allowing them to be used as pulling devices. While the ram can be used for pulling, steps should be taken to avoid damaging the tool. Easy comparisons can be made between a fully extended ram ready to perform a pulling operation, and a ladder at full extension. In both cases, the tool is at its weakest configuration and susceptible to damage if a force is exerted laterally in the middle of the device. To avoid damaging the tool if used for a pulling operation, the ram should not make contact with anything between the two chain connecting points.

An example of when a ram was used successfully in a pulling operation was a train versus car crash. The driver of the car was trapped, necessitating the relocation of the steering column. In this case, one chain was attached to the steering column and the other chain was connected to the locomotive. The ram was free of any side loading, allowing the ram to make the pull without any damage to the tool.

Other attachments for use with spreaders and rams include the following:

- Base plates that can be attached to spreaders or rams to create a wider, more stable base
- V blocks for replacement of the standard ram head, when extra gripping is required

Hose reels can reduce set-up time and are helpful when power plants are very heavy. Using reels allows the most frequently used tools to be stored while still connected to the power plant; this reduces set-up time. Once the unit has arrived at the scene, crew members simply pick up the tool and walk to the area where the tool is needed, while the driver/operator feeds the hose and gets the power plant up and running. This set up is especially valuable when the apparatus is outfitted with a power plant known as a simo-pump. A simo-pump is equipped with two or four selector valves that allow two to four tools to be hooked up and working at the same time without loss of pressure.

Another benefit of having the tools hooked up on a reel system is when the apparatus won't be the first- or second-arriving unit at the crash scene and won't get first choice in positioning at the scene. Typical reels carry either 50 or 100 ft of hose. This allows the hoses to be stretched to the work area—despite the apparatus positioning at the scene. If, however, the hoses won't reach the work area, the power plant can be disconnected from the reels and carried to the scene. For this situation, it's a good idea to always have two or three spare lengths of hose on the apparatus to use in place of the reel.

If an apparatus is to be fitted with more than one hydraulic hose reel, given a choice, the reels should be outfitted with different color hoses. Working and moving around a crash scene can cause hoses to cross over one another and even become tangled. If the hoses are the same color and one of the tools must be shut down/dumped, the crew member at the power plant simply has to know the color of the hose attached to the tool and then can operate the controls at the power plant without any confusion.

When more than one tool is being attached to a power plant, a manifold block simplifies the hose hookup procedure. Manifold blocks are designed to provide a continuous flow of hydraulic fluid from one tool to the next and then back to the power plant. Some manifold blocks are equipped with dump valves that allow a crew member at the manifold block to control the flow of hydraulic fluid. The dump valve at the manifold block works the same way as the dump valve that's mounted on the power plant. This is a useful accessory when the tools will be working off a reel and operating a distance from the power plant dump valve.

Modern spreaders, cutters, and rams

Since the introduction of the first hydraulic rescue tools, significant advancements have been made in the weight, size, and power of the tools. As the market for the tools expanded, so has the number of companies manufacturing and selling them. While the tools have many common operating characteristics, there are also many differences. For this reason, and in the interest of obtaining maximum effectiveness with the tools, the manufacturers' representatives should be considered the best sources for detailed instructions on the proper use of the tools.

There are, however, some general guidelines that apply to all tools.

- Proper PPE, especially eye protection, is important. Despite advances in engineering and materials, hoses still burst on occasion, just like fire hoses, and objects fly off of vehicles when they're cut. Simple eye protection can help prevent a premature, miserable, line-of-duty disability retirement. It's just that simple.

- Good body mechanics can prevent fatigue. Today's spreaders are half the weight they were in the 1970s but can still be fatiguing during long operations. Good body mechanics can reduce fatigue to the shoulders and arms and reduce the chances of a back injury.

- Stay out of the way of moving parts of both the tools and the vehicle. When selecting an approach with a tool, the operator should anticipate how the tool and the vehicle will respond to the action of the tool. The tool operator should also watch out for other crew members who are working in the area of the hydraulic tools.

- When the tools are being operated, nobody should have their hands on the moving parts of the tools, including the spreader arms, cutter blades, and ram pistons. On occasion, it may be necessary to support a ram in an awkward position until it gets a grip on the vehicle, or rotate the ram to get the ram head in a proper position. These situations should be the exception, not the norm, when operating the tools. Hydraulic tools are unforgiving when it comes to carelessness, and their power should be respected.

- As the tools have become more powerful, greater responsibility for their appropriate use has fallen on the tool operators. Older, lower power tools almost never broke cutter blades or spreader arms, but that seems to be changing. Tools are being damaged by operators and officers that are making poor decisions on tool application. To reduce the damage to cutters, the tool operator should watch out for blades that are separating, and should use the deepest part of the cutter when positioning the tool for a cut. Attempts to punch through heavy steel with the cutter tips are one of the most common causes of blade fractures.

Power plants— a closer look

Power plants are the backbone of any heavy hydraulic system, and crew members using the equipment should have a good understanding of how they work, how to use them, and how to take care of them. By understanding how they work, problems can be avoided or resolved on the crash scene.

A power plant has a few major components.

- A source of power, such as an engine, an electric motor, or a power take off (PTO)
- A pump that is attached to the power source
- A reservoir tank that holds the hydraulic fluid
- The plumbing and valves needed to put it all together

Power source. Modern hydraulic power plants can be fitted with a variety of power sources to meet the needs of the users. The following power sources can be outfitted to drive a hydraulic pump:

- Four-stroke engine
- Electric motor
- Two-stroke engine
- Diesel engine
- PTO

The most common power source today for heavy hydraulic pumps is the four-stroke engine. The four-stroke engine has virtually replaced the two-stroke engine as a power source because it is considered less temperamental and requires less maintenance and attention. That doesn't mean the four-stroke is a "gas and forget it" engine; it just seems to do better with the type of use the fire service gives it. For those new to engines, extrication tools, or the fire service in general, it's helpful to understand what makes a four-stroke a four-stroke, and how it can be differentiated from a two-stroke engine.

Most engines operate basically the same way. To develop power, a piston goes up and down inside a cylinder. The cylinder and the piston are designed to have a snug fit but still allow for the movement of the piston. The piston is connected to a part called a *crankshaft* by means of a connecting rod. As the piston moves up and down in the cylinder, it causes the crankshaft to spin. The faster the piston moves up and down in the cylinder, the faster the crankshaft spins. This arrangement allows the up and down motion of the piston to be converted to the spinning motion of the crankshaft. The spinning crankshaft is the part that provides power for virtually all machines that use internal combustion engines for a power source.

To make the piston move up and down in the cylinder, a mist of gasoline and air is brought into the cylinder by a valve located above the piston. As the piston moves down in the cylinder, the mist is drawn into the cylinder above the piston. This is called the *intake stroke*. This air/gas mixture is created by the carburetor—a part that is located on the outside of the engine.

Vehicle Extrication: A Practical Guide

Once the gas/air mixture is in the cylinder above the piston, the piston moves up to compress it. This is called the *compression stroke*. With the gas/air mixture under pressure and the piston at about the top of its rotation in the cylinder, a spark plug, which is also located above the piston near the valves, causes a small electric arc to occur in the gas/air mist. When the arc ignites the mist, the piston is propelled down the cylinder. The small explosion caused by the ignition of the mist causes the piston to move down quickly, which causes the crankshaft to spin. This is called the *power stroke*.

A typical four-stroke engine used on many powerplants.

As the piston moves up and down in the cylinder, the crankshaft spins.

As the crankshaft spins, a dipper picks up oil to lubricate the engine.

The crankshaft fits into the center of this horizontally mounted pump.

Fig. 2–23 Powerplant components—four stroke engine and pump

Tools, Equipment, and Apparatus

After the piston reaches the bottom of cylinder and starts back up again, another valve, known as the exhaust valve, opens up at the top of the cylinder. As the piston moves up in the cylinder, instead of compressing more gas/air mist as it did during the compression stroke, it forces the burnt mixture out the open exhaust valve. This is called the *exhaust stroke*. The cylinder is then clear of any burnt fuel and is ready to take in more gas/air mist during the next intake stroke. This occurs after the piston reaches the top of the cylinder and starts down to start the whole cycle again.

Side view of the engine.

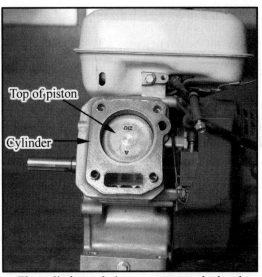

The cylinder and piston connect to the head.

The head includes the intake and exhaust valves and the bottom part of the spark plug.

The major components assembled.

Fig. 2–24 Powerplant components—four stroke engine

The term *four-stroke engine* is used to describe this type of engine because there are four distinct strokes of activity: intake, compression, power, and exhaust. When the piston is moving up in the cylinder, the engine is in either the compression or exhaust phases, and when the piston is moving down it's in either the power or intake phase. Two-stroke engines operate in a different manner, combining strokes, thus the name.

Engine lubrication. With all this movement going on in the hot environment inside an engine, the engine needs some kind of lubrication or it will overheat and seize up. To supply the needed lubrication, oil is stored in the bottom of the engine in a section called the *sump*. As the engine runs, the oil is used to lubricate the internal parts and dissipate heat; some oil will be consumed and will need to be replaced in order to maintain adequate lubrication. The more the engine is used and worn, the looser the fit between the internal parts; this results in higher oil consumption. To make sure that the engine has the right amount of oil to lubricate the engine, a dipstick or other indicator is often used to check the amount of oil in the oil sump. Some manufacturers may have different instructions for checking the oil level, but the most common procedures are described in the following paragraphs.

Checking the oil in an engine is a simple procedure, but as always, the manufacturer's owner's manual should be referenced for specific procedures. Following is a general guide for engines equipped with a dipstick:

1. Unscrew the dipstick and remove it from the engine. If the engine has been running, allow the engine to cool down for a few minutes so the oil can flow down to the sump to ensure an accurate measurement.

2. Wipe the dipstick clean with a rag.

3. Reinsert the dipstick into the engine without screwing it in. When reinserting the dipstick, avoid rubbing the dipstick against the internal engine parts, because rubbing oil on the dipstick that could result in a false reading.

4. Remove and read the dipstick. Usually there are markings on the dipstick that show the appropriate oil level. The markings may include the following abbreviations:

    - Min (minimum level that should be maintained).

    - Max (maximum level that should be used).

    - Add (oil should be added).

    - OK (the oil level is in the acceptable range).

    - Sometimes there is simply a cross-hatched area on the dipstick indicating the proper range.

As engine designs change, so do the methods of maintaining them. For this reason, follow the owner's manuals for the best source of information about oil and fuel levels. To help owners avoid damage, some of the newer engine designs incorporate an automatic shutoff when the oil level falls below acceptable levels.

Fuel. Checking the fuel in a power plant is usually pretty simple. The fuel cap is removed and the level is observed. It is usually best to keep the fuel level at a point where it's close to full but not so full that it leaks out when the apparatus is driven around town. Following are a couple of

Tools, Equipment, and Apparatus

fuel problems that prevent an engine from operating properly:

- Contaminated fuel—this type of problem is usually caused by rust or water. The rust occurs when old metal fuel storage tanks are used to store fuel. The rust is mixed with the fuel and is poured into the engine. The water is caused by condensation that occurs in partially filled fuel containers. Either contaminant will cause an engine to stop operating properly.

- Stale fuel—fuel that has been stored for a long time changes into substances that cause an engine to either run poorly or not at all. The key here is to develop a fuel rotation program so that tools that are used infrequently have the fuel removed and replaced with fresh fuel. Regular operating of the equipment keeps fresh fuel moving through the engine, which reduces the chances of engine malfunction.

- Wrong fuel—inadequately marked fuel containers can result in the wrong fuels being used in engines. Diesel, gas, and mixed fuel containers should be marked properly to avoid problems. If an incorrect fuel is used in an engine, the results can range from an engine that won't run to an engine that is destroyed for lack of lubrication.

Which fuel to use? Sometimes crew members encounter unfamiliar engines that need fuel. The question then becomes, *which fuel do I use?* All of the obvious resources should be used to determine the correct answer to the question, but it's also nice to know how others can just look at an engine and tell what kind of fuel is needed. While there are no hard and fast rules that can be applied in all situations, there are a couple of big clues that can be used.

- Does the engine have spark plugs and spark plug wires? If there are no spark plugs, chances are good that it's a diesel engine. Some manufacturers make small diesel engines available for their power plants.

- Is there a place to check the oil level in the engine? If there's a dipstick or plug for checking the engine oil, chances are good that it's a four-stroke engine. Most small two-stroke engines use fuel that has two-stroke oil added to it. This is done because two-stroke engines, by design, don't use an engine oil reservoir or sump. Some two-stroke engines, like those used in some outboard boat engines, use a separate oil tank. In these engines, the oil is stored in a separate oil tank and is then injected into the fuel as it's pumped into the engine. When checking the engine oil, it's important not to confuse the engine oil level with the hydraulic reservoir oil levels.

The pump and hydraulic reservoir. The pump receives its power from the crankshaft. The pump can be located either beneath the engine or to the side of the engine. This is determined by the type of engine that is being used to power the pump. In either case the pump is submerged in hydraulic fluid that is used to operate the tools. Manufacturers use a few different pumps, but they all do the same thing: pump fluid. Following are important differences:

- Most pumps develop either 5000, 10,000, or 10,500 psi.

- There are two common types of hydraulic fluid used: phosphate ester and mineral oil.

- Pumps, hoses, and tools are fitted with O rings that are compatible with the fluid being used.

Fig. 2–25 Hydraulic Fluid. Two types of hydraulic fluid commonly used in hydraulic rescue tools: Phosphate Ester on the left, Mineral Oil on the right.

Knowing about the fluids and pressures that tools use is important for the following reasons:

- If given a choice between two containers of hydraulic fluid, a crew member should know which fluid is used in a particular tool system before adding the fluid.

- If given no choice, but handed a container of fluid, the crew member should know if it's the wrong fluid and not use it.

- If a crew member gets blasted in the face with hydraulic fluid from a burst hose and needs to be seen by a physician, it's nice to be able to say what inflicted the injury.

- It's a sign of professionalism. It's like knowing the size of the pump and water tank on the apparatus.

Starting the power plant. As with most tool systems, it's best to build the entire system before putting it into operation. Once a crew is familiar with a heavy hydraulic system, changes in the basic procedure can be made in the interest of efficiency. For example, if a crew is using a power plant that can supply and control two tools independently of one another (simo-pump), the officer may have one crew member start working with a spreader, while the cutter tool and other hoses are set up. By not building the entire system before putting it into operation, one tool can start working faster, but this also increases the chance of a discharge or hose being charged without a tool being hooked up. Experience and training can help the officer make the best decision on how to start the operations.

When a crew member is told to crank up the power plant, developing a routine can help prevent problems. A typical start-up routine can include the following steps.

1. Place the power plant in the best position for the tool operators.
2. Place the power plant on a level surface, or as level as possible.
3. Connect the hoses to the tools.
4. Connect the hoses to the power plant.
5. Ensure the dump/selector valve is in dump or neutral position.
6. Open the fuel valve.
7. Switch the ignition switch to the "on" position.
8. Move the throttle lever to the run/start/fast position. Engines are different, but the throttle probably will be marked with one of these settings.
9. Turn on the choke (if equipped).

Tools, Equipment, and Apparatus

10. Operate fuel pump plunger (if equipped).
11. Pull recoil starter cord a couple of times, until engine starts, or sputters and stalls.
12. Turn off the choke.
13. Pull the recoil starter cord again (if engine stalled during first attempt to start).
14. Move the dump/selector valve to the "pressure" or selected position.

Does the engine have a spark plug? Gasoline engines use spark plugs.

This engine doesn't have spark plugs. This is a diesel engine.

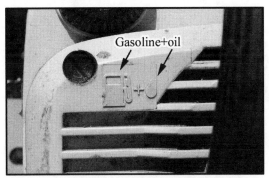

This saw uses a gas/oil mix for fuel. This identifies it as a two-stroke engine.

The oil reservoir on this two-stroke chain saw is for lubricating the chain, not the engine.

The presence of an oil dip stick indicates the engine uses straight gasoline.

The best clue is the label or marking on the engine.

Fig. 2–26 Which fuel to use? Gas, diesel, or gas/oil mix? Clues to help select the right fuel.

Steps involving checking the levels of engine oil, fuel, and hydraulic fluid typically are not part of the start-up procedure on the crash scene because these items should have been checked as part of the routine daily equipment checkout at the station.

Shutting the power plant down. To end operations, use the following steps as a guide:

- Open or close the tools to the position desired for storage on the apparatus. Usually spreaders are closed completely

The right half of this powerplant consists of the hydraulic pump and reservoir.

A dip stick is used to check the level of hydraulic fluid in the reservoir.

The dip stick is removed, wiped clean, then reinserted into the tank.

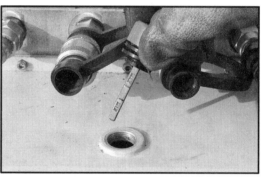

This dip stick is marked with "MIN" and "MAX" levels.

This powerplant uses a top mounted gauge. Levels are marked in red, yellow, and green.

This side mounted sight gauge also shows the level of fluid in the reservoir.

Fig. 2–27 Powerplant components—the hydraulic fluid reservoir

then opened about 1/2 in. Cutters are closed so that the tips of each blade override one another to keep the tips from snagging. Rams are retracted to the fully closed position and then extended about 1/2 in. It's preferable to store the tools slightly opened/extended to relieve any hydraulic pressure in the tool.

- Move the dump/selector valve(s) to the dump/neutral position.

- Move the throttle to the idle or off position.

- Turn the ignition switch to the "off" position (if equipped).

- Disconnect, clean as necessary, and recoil the hoses for storage.

Returning to service. After the engine has been shut off it should be cooled for a while, then refueled. If there were any problems with hose or tool leaks during the operation, the hydraulic fluid level should be checked and serviced as needed.

the purchasers. The time to implement improvements in tool layout is during the needs assessment and design phase of production.

Items to be addressed include the mounting of power plants, tools, reels, small hand tools, and lighting. Manufacturers are constantly improving the functionality of the rigs, often based on the recommendations of their clients. Solutions to some of the most common problems are illustrated in the following section.

Apparatus

The way in which extrication tools are stored on an apparatus has a big impact on the way the crew functions when they arrive on a crash scene. Equipment that is thoughtfully laid out and easily deployed gives the crew the ability to work quickly and efficiently at the crash scene. While many of the design features that go into building apparatus are addressed in National Fire Protection Association (NFPA) standards, many design options are left to the manufacturer and

Vehicle Extrication: A Practical Guide

APPENDIX:
APPARATUS FEATURES AND TOOL STORAGE

PTO generators are becoming common features on new special operations units.

These generators are started from the cab by simply pushing a button.

PTO generators can be mounted in areas that free up compartment space.

The generator when viewed from below the running boards and compartments.

With plenty of power available, this unit was outfitted with 240 volt outlets for future use.

The rating plate for the generator is mounted on the cab for easy reference.

Fig. 2–28 PTO generators—providing plenty of power for electric extrication tools

Tools, Equipment, and Apparatus

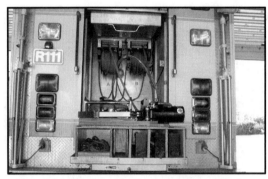

Dual hydraulic reels are preconnected to the tools; rams and chains are stored below.

Instead of reels, two small powerplants are preconnected to the tools.

Lightweight tools are mounted on the doors of this extrication compartment.

Tool brackets can enhance access to the tools and make good use of all compartment space.

The top of this compartment is outfitted with reels and tools.

Portable powerplants provide flexibility in the method of deployment.

Fig. 2–29 Hydraulic tool storage—different approaches to optimize storage and deployment of tools

Vehicle Extrication: A Practical Guide

To access this high compartment, the drawer is first pulled out.

The drawer then pivots downward.

By pivoting downward, better access and a better view is obtained.

Injuries can be reduced by providing good access to awkward and heavy tools.

This pass-through compartment holds long 4 x 4 and 6 x 6 material.

Low, underutilized areas can provide good storage possibilities for cribbing.

Fig. 2–30 Compartment options—improved access to heavy and bulky tools

Tools, Equipment, and Apparatus

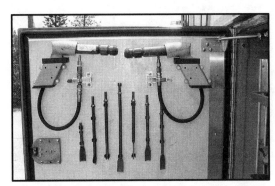
Mounting these air chisels on the apparatus door makes bit selection quick and easy.

Cloth chain bags are easy to carry and store; an improvement over storing them loose on a shelf.

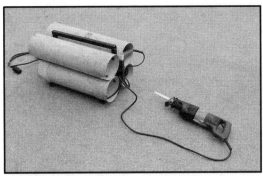
This equipment carrier allows two saws to be put in service quickly.

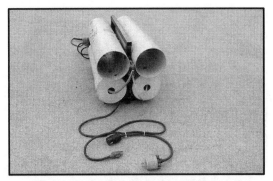
The carrier is equipped with adaptors and and extension cords.

Instead of carrying cribbing like fire wood, straps allow this cribbing to be carried in one hand.

To release this bundle, the bight is slipped over the knot.

Fig. 2–31 Storage options—little things that make the job easier

Vehicle Extrication: A Practical Guide

Apparatus-mounted reels are the quickest, most efficient way to get power to the work area.

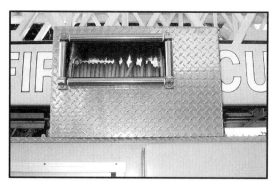

Reels are often mounted high, in locations that would otherwise go unused.

If mounted reels are out of the question, portable reels may provide satisfactory service.

Whichever type of equipment is used, compatibility of the plugs is a critical issue.

This modern box has the amperage rating on each door.

Simple plastic reels, though not preconnected, provide for quick tool set up.

Fig. 2–32 Electric power distribution—getting power to the scene

Tools, Equipment, and Apparatus

New apparatus are coming equipped with significant improvements in scene lighting.

This engine is equipped with multiple lighting fixtures for a variety of scene needs.

These lights provide coverage for three areas of the scene.

This floodlight telescopes, pivots, and rotates.

For good scene and work area coverage, lights are provided on all four sides of the apparatus.

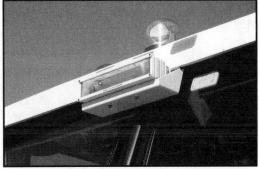
This flood is mounted at the top center of the windshield for when the apparatus is nose-in.

Fig. 2–33 Enhanced apparatus lights—providing for improved scene lighting

Vehicle Extrication: A Practical Guide

Pneumatic telescoping mast lights provide up to 12,000 watts of light almost instantly.

With their simple set up, large crash scenes can be easily illuminated by one crew member.

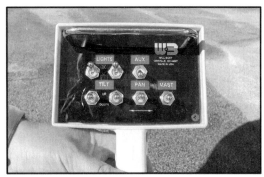

The remote control gives the operator control without climbing up on the apparatus.

With no batteries to wear down, a heavy duty droplight is an option for long operations.

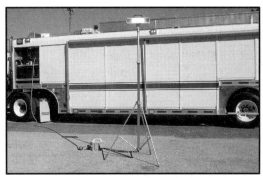

These multi-function lights can be used on a tripod.

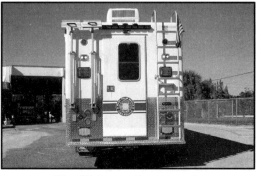

Floodlights can be operated while in their mounting brackets.

Fig. 2–34 Scene lighting—telescoping and portable systems

Tools, Equipment, and Apparatus

Companies that run on a lot of crashes may use a carpenter's belt to keep tools organized.

Shears for cutting cloth and belts, a center punch for tempered glass and a utility knife.

Electrician's pliers, Channel-locks, fence pliers, and wire cutters are useful for small prying jobs.

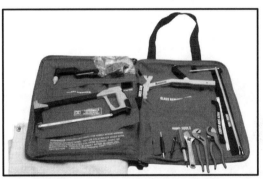
This kit by Howell Rescue contains all the small tools typically used at a crash scene.

A loop of inner tube can be used to keep shears within reach when belts must be cut.

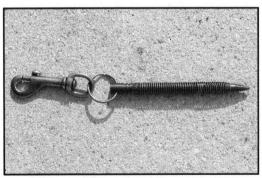
There are a variety of clips that can make center punches easy to find.

Fig. 2–35 Small hand tools—storing and carrying for quick access

· 3 ·

Stabilization

Stabilization discussions usually include ways of using cribbing to prevent a vehicle from moving as firefighters perform extrication procedures. Stabilization of the vehicle is important, but there are other important concerns at a crash scene that also need to be addressed. Stabilization should be addressed in the following three distinct, but often overlapping, areas:

- Scene (perimeter and core)
- Vehicle
- Occupants

The typical sequence of stabilization actions is based on the sequence of contact at the crash. Stabilize the scene first, followed by the vehicle, and then the occupants. The sequence is typical but not cast in stone; each crash incident should be analyzed to determine the most appropriate sequence. Tactics used in stabilization should be a result of size-up by the company officer, who must determine the most effective methods needed to bring the incident to a predictable, successful conclusion.

STABILIZING THE SCENE

The responsibility for getting the crash scene under control or stabilized is usually shared by law enforcement and the fire department. Law enforcement officers close off roadways and control bystanders by establishing and maintaining a perimeter around the scene. This perimeter allows firefighters to concentrate on tasks associated with providing occupant rescue and treatment. There are times, however, when there are an insufficient number of law enforcement officers on the scene to adequately stabilize the perimeter of the crash scene. In these cases, the fire officer should address the needs of the perimeter in addition to the core of the scene.

The 911 call indicated this vehicle's occupants were trapped and the vehicle was burning.

One side of the intersection (scene) is stabilized by blocking the scene with apparatus.

Police units assist in stabilizing the scene by blocking another side of the intersection.

Stabilizing the vehicle required extinguishment of the engine compartment fire.

To access the engine compartment completely, a saw was used to cut through the hood latch.

Stabilization of the victims includes finding all the injured victims that fled the burning vehicle.

Fig. 3–1 Stabilization—scene, vehicle, and victims

Stabilization

Effective deployment of equipment and personnel at a crash scene should occur after the company officer has performed a risk/benefit analysis of possible actions at the crash scene. To perform this analysis, the officer needs to have a clear understanding of the risks before any benefit analysis can be considered.

Consider the following scenario: an engine and a medic unit are dispatched at 2:00 AM to a reported crash on a busy section of an interstate highway. As the units approach the area, it is evident that the crash is on the far or downhill side of an overpass. Being on the downhill side means that drivers approaching the scene at 70 miles per hour will not have enough time to recognize and respond to what they are seeing as they approach the scene. How should the officer begin stabilizing the scene? Which is of greater importance, establishing a perimeter, which means blocking traffic from entering the crash scene, or assisting the medic unit on the scene? The officer should quickly perform a risk/benefit analysis and then take appropriate action.

In this scenario there would be significant risk to everyone who is working at the crash scene if the traffic isn't slowed and redirected at the top of the hill. One course of action is to have the medic unit proceed to the scene while the engine blocks traffic at the top of the overpass. By placing the engine with its reflecting striping, warning lights and large mass at the top of the overpass or hill, the traffic approaching the area will have a longer period of time to react to the situation, which could be critical at 2:00 AM. Personnel on the engine can place fusees (flares) and cones on the highway to provide additional visual warning and then can walk to the scene if needed. It's a good idea to leave the driver of the apparatus with the engine so that it can be brought to the scene when law enforcement personnel assume responsibility for the scene perimeter.

Upon arrival at the scene core, the company officer may determine that the engine can be best used at that location and may direct the driver to relocate the apparatus. It's important that all personnel pay attention to traffic at all times to avoid being struck by drivers who are confused by the activity or are impaired. It is all too common for law enforcement officers and firefighters to be struck at crash scenes, reinforcing the need for a high level of situational awareness.

There are situations when the units arrive at the scene from one direction but the area that needs to be blocked is on the other side of the crash scene. In this situation, the officer can consider dropping personnel and equipment at the scene and having the driver proceed to the area needed.

This procedure is similar to making a reverse lay with fire hose; it's important to get everything off the apparatus that will be needed before it pulls away from the scene.

Whether the apparatus is parked at the scene perimeter to protect the scene from oncoming traffic or at the scene core, a decision must be made about which lane or lanes should be blocked. This is a subject that, over the years, has put firefighters and law enforcement personnel at odds, and in extreme cases has resulted in arrests. As with many differences in opinion, often, the conflict is a result of lack of communication and understanding between the parties involved in management of the scene. Law enforcement personnel have a responsibility to open as many

Vehicle Extrication: A Practical Guide

lanes and roads as quickly as possible and to minimize delays to the public. This may seem like a trivial concern, but closed lanes and roads can have a significant financial impact on the community and its businesses. Additionally, as drivers gawk and travel around a crash scene someone invariably causes another crash, and the crash cycle repeats itself.

Traffic (scene) control is established by use of warning lights, cones and mass.

When struck, cones produce a distinctive sound, alerting crews to rapidly approaching danger.

If traffic cones are carried on apparatus, those with reflective striping should be considered.

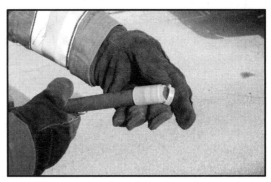

If flares are used at night, the striker cap is first removed.

The striker is then removed, reversed, and used to strike the end of the flare to ignite it.

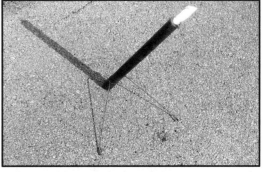

The metal legs can be bent down to support and elevate the flare.

Fig. 3–2 Scene stabilization—warning devices for motorists and crews

Stabilization

Protecting the scene with effective apparatus placement

If the officer elects to respond to the crash core and decides that one or two lanes of travel should be closed to protect personnel, the driver can approach the scene in one lane and then angle the apparatus into the second lane prior to stopping the apparatus. How far to stop the apparatus from the vehicles can be based on several factors.

- Deployment of personnel—the apparatus can be parked so that personnel can reach the vehicles quickly as they step off the apparatus.

- Deployment of equipment—hydraulic tools are heavy, making it advantageous to park the apparatus close to the vehicles. If tools are set up to run off of electric, pneumatic, or hydraulic reels, the apparatus should be spotted so the cords or hoses on the reels can reach the vehicles.

- Lighting—if the apparatus scene lights are to be used, the apparatus should be positioned close enough for the scene lights to be effective.

- Hose deployment—coming up short with fire hose on a crash scene is uncommon, but excessive hose can create trip hazards for personnel operating on the scene. An approach that has become popular is to pull the front bumper trash line. This line is usually a 100-ft length of 1-3/4 in. hose. These short stretches result in less excessive hose and quicker, easier reloading times, which translate into a greater likelihood of deployment.

- Leaking fluids—if the vehicles in the crash are leaking fluids, especially flammable liquids, the apparatus should be parked so that the fluids don't flow under the apparatus. This may necessitate parking uphill and/or upwind. Basic HAZMAT operational awareness helps here.

- Danger from approaching vehicles—if traffic is typically traveling at a high rate of speed, anticipate how the apparatus would move if struck by another vehicle. The idea is to keep the apparatus from sliding into the work area if struck by another vehicle.

Fig. 3–3 Scene lighting was provided by the light on the end of this 50-ft Squrt.

Drivers and officers may be hesitant to park their apparatus in a position that could result in being struck by other vehicles. Often this concern is based on experience that tells them that any damage to the apparatus will result in a massive amount of paperwork, photos, and investigations. Once parked to protect a scene, an apparatus driver may be tempted to reposition the apparatus based on the amount of tire-squealing being produced by approaching vehicles. This is a time when clear thinking will lead to good outcomes by recognizing the following:

- If there is a lot of tire-squealing heard, there is probably insufficient reaction time/distance established for the drivers of vehicles approaching the scene. Earlier warning is needed to allow drivers the opportunity to slow down in a more controlled manner.

- You can replace the apparatus, but you can't replace people. If the mental image of a car striking an apparatus is bad, imagine the same car striking a crew working on the scene. It is simply unacceptable to sacrifice flesh and bone for the preservation of steel and aluminum.

- If law enforcement personnel aren't on the scene in numbers large enough to create a safe work environment, the officer should consider calling for an additional fire unit to help protect the scene and crew.

One very dangerous crash situation may remain unacknowledged until the quiet of the scene is broken by the air horns of a train locomotive. Whenever a vehicle is found disabled on railroad tracks, especially with its occupants trapped, the local railroad office should be immediately notified so trains heading into the area can be stopped before approaching the crash scene.

Fig. 3–4 Railroad authorities should be notified immediately of any crash occurring on railroad tracks.

In rural areas, or any other area where radio and cellular phone service isn't available, or if the railroad office can't be contacted, the officer should provide adequate warning for oncoming trains so they can stop before approaching the scene. Local railroad companies should be contacted to learn about local policies and procedures for stopping a train in case of emergency.

Fig. 3–5 Signs located at railroad crossings provide emergency personnel with the information needed when contacting the railroad about vehicles stopped on the tracks. The signs may include the phone number to call, the crossing number, and the railroad mileage number.

Stabilization

Identifying utility hazards at the scene core

Traffic and trains are not the only hazards that can be present on a crash scene. Utility equipment that is damaged by vehicles involved in a crash can cause a very serious safety concern for all operating at a scene. Probably one of the most dangerous situations is when a vehicle causes damage to an electrical utility device. Utility poles that are damaged are pretty easy to spot, but the wires hanging from the poles can be difficult to see—especially at night (see Fig. 3–6.) This situation can result in apparatus driving into or over the wires if they're draping

A light pole lies along a fence after being struck. The question is: "Where's the base?"

The spacing and location of other lights indicate the pole has been knocked away from its base.

From a distance of less than ten feet, it's still difficult to locate the base and wires.

Without a good size-up, this base could remain undetected beneath a vehicle.

Fortunately this light has been fitted with a plug designed to reduce the danger of exposed wires.

Even with this improvement, there's still a risk of torn insulation and exposed wires.

Fig. 3–6 Scene stabilization—where is the hidden electrical hazard?

across a street. If the pole is broken but the wires are intact the wires can be stretched tight, leaving the upper portion of the pole suspended in the air. If the load is too great for the wires they will break and fall onto the scene, possibly onto the vehicles involved in the crash.

Not all electrical problems involve overhead power lines, however. It's easy to miss the remains of damaged equipment that is powered by underground wires during size-up because of its small size and tendency to blend into the scene.

To reduce the chance of confusion, crews should determine which agency is responsible for electrical utility equipment in their response area. It wouldn't be surprising to find that a municipality is responsible for some light poles while others are operated by the local Department of Transportation or local power company. In addition to light poles, other devices that can receive electrical service from underground lines include the following:

- Transformers
- Utility poles
- Landscape and architectural lighting
- Traffic control devices
- Bus stop benches

These devices are particularly dangerous because when struck by a vehicle the underground equipment cover or attachment can fly off the device, exposing bare wires. If the vehicle comes to rest on top of the exposed wires, the entire vehicle can become energized, yet the danger may not be obvious. This is a situation that has been known to kill firefighters and should be dealt with before anybody touches the vehicles.

Fig. 3–7 A good size-up and walk-around is needed to identify all potential hazards and victims.

To identify most electrical hazards on a scene, the company officer can take the following steps:

- Observe the typical spacing of light and power poles as the apparatus approaches the scene. A gap in the typical sequence may be an indication that one of the poles has been knocked down.

- Use a spotlight to look for poles and wires that have been partially knocked down and pose a risk to the crew when approaching the scene in the dark.

- Scan the scene to find any poles or utility boxes lying on the ground. The next step is to locate the base of the device. Once the base has been located, the area can be marked with scene tape, traffic cones, or personnel may be used to prevent anyone from accidentally coming in contact with the wires.

- Remember when approaching the vehicle there are three sides of the vehicle that probably have not been seen: the underside and the two sides on the far side of the vehicle. This means that 50% of the sides haven't been seen as the crew members walk up to the vehicle. These areas should be observed before crew members touch the vehicles.

Whenever wires are found on the ground or out of their normal position, they should be treated as live until a power company representative advises the crew otherwise.

Vehicles in contact with live wires

Oddly enough, there are no universal guides for action in all parts of the country according to power company officials. Each power company and responding agency has different needs and problems necessitating different responses. Without knowledge of any specific guide to action, the most conservative, safest action should be taken: call the power company for help. Some companies like Florida Power and Light (FPL) have developed excellent training packages for Florida firefighters to help expedite the power company's response to an emergency. The Job Aids package that FPL has produced not only provides basic guidelines for safe operations, but it also provides a method of identifying and describing the type of equipment that has been damaged at the scene. With a good size-up of the problem and clear communication with the power company, the correct type of assistance can be sent to the scene. To determine if training packages and procedures have been developed by local power companies, contact local power company safety office representatives.

STABILIZATION OF THE VEHICLE

Shutting down the vehicle

Once the scene has been stabilized, attention can be given to the vehicle. There will be situations when it is best to work on stabilizing the scene and vehicle simultaneously if adequate personnel are present. This can reduce the time needed to perform any rescue procedures that are required to free the occupants of the vehicle. Stabilizing a vehicle does not always include some exotic cribbing and strapping procedure but, in fact, can be very simple.

The following universal steps that can be taken to stabilize a vehicle:

- Chock the tires. This is especially important when the vehicle is on a hill.

- Set the parking brake. There is no downside to setting the parking brake, and it can be reversed at a later time if necessary.

- If the engine is running, shut it down by turning the key. By turning the ignition off, the risk of an accidental air bag deployment is decreased.

- Shift the transmission into "park."

- Disconnect the battery or batteries. Many diesel engines utilize two batteries. This important step reduces the chance of an electrical fire and helps minimize the risk posed by air bag systems. If the only step taken to disconnect the electrical supply is

Vehicle Extrication: A Practical Guide

to turn the ignition key to the "off" position, power may remain intact if a short circuit occurs during an extrication procedure.

- Crib as appropriate.

The order of steps is different than the way vehicles are normally stopped and turned off. Training manuals have historically recommended that the vehicle first be placed in "park" followed by the ignition being turned off. Consider what would happen if the front end of a vehicle with

Chocking the wheels.

Setting the parking brake.

Turning the ignition off and removing the key.

Shifting the transmission to park.

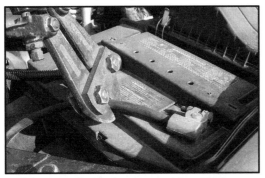
Disconnecting the battery, negative side first.

Using appropriate cribbing to provide a stable vehicle that's safe for victims and crewmembers.

Fig. 3–8 Vehicle stabilization—simple steps for a safe operation

the engine running and in gear were pressed against another vehicle after a crash. In the case of an automatic transmission, what would happen if an attempt was made to shift the transmission into "park" from outside the vehicle? The crew member doing this would be outside the vehicle, without putting any pressure on the brake pedal. As the selector slides past "neutral" on the way to "park" it must pass "reverse." If the transmission linkage is damaged, the shift lever could stop in the "reverse" position before reaching the "park" position. If the engine is running and the vehicle is placed in reverse, the vehicle could start moving rearward under power.

In this type of situation, local protocol and policy should serve as a guide to action. While the sequence of turning a vehicle off seems like a subject too small for serious consideration, the consequences of a misstep could have significant implications at a crash scene.

The hybrids

Feeling the hood and looking at the dashboard to determine if an engine is running may be a little confusing if one of the vehicles involved in the crash is a hybrid. The term *hybrid* means that two variations of an object are combined to make a new type of object. When hybrid is applied to automobiles it means that the vehicle uses both an internal combustion engine and an electric motor to power the vehicle. When sitting idle, the engine of a hybrid may not be running at all, which can cause some confusion for a crew member not familiar with identifying a hybrid vehicle. The good news is that the steps for shutting down a hybrid are basically the same as a conventional automobile.

Until recently, hybrid vehicles were considered the car of the future and only seen in auto magazines.

The hybrids on the street today remain anonymous because of the lack of distinguishing styling details. To help firefighters recognize and understand these new forms of vehicles, companies like Toyota have published emergency response guides for their product lines that explain how the vehicles work and how their design impacts firefighters. These guides, along with websites and magazine articles, are the best way to stay on top of the continuous changes in engineering and design of all types of vehicles.

For most, memorizing all the details of every model of automobile to limit risk while working on a vehicle would be an impossible and impractical task. At the crash scene, it's difficult enough to identify the manufacturer of an automobile, let alone the model year and construction details. One practical approach to this problem is to have a good understanding of how most vehicles operate and what makes the unusual ones different. The Toyota Prius in the following photographs has more in common with standard vehicles than not (See Fig.3–9). It has four doors, four tires, two bucket seats, and a gasoline engine. The big things that make it different are the electric motor, power cables, and the high-voltage battery pack in the trunk. Knowing that these are the things that make the vehicle different, the question to be answered is *what makes a difference to firefighters on the scene?*

The answer to this question can be determined by looking at how the hybrids are built and then considering how they would sustain a crash. By thinking about crash scenes involving non-hybrids, and mentally putting a hybrid in its

Vehicle Extrication: A Practical Guide

place, the crew can visualize potential problems that will need to be resolved. Fortunately, the manufacturers look at things the same way and design systems and parts that minimize the hazards to passengers and firefighters. When these new types of vehicles are involved in crashes, the designers and engineers will look carefully at the results and make improvements in their crash-worthiness. The key to effective operations is to be familiar with all types of vehicles to avoid unnecessary apprehension and tactical paralysis.

Hybrids are on the street today and may look like other traditionally powered vehicles.

One way to identify a hybrid is by the distinctive emblems.

When shutting down a hybrid, the first difference observed may be the dashboard.

The Energy Monitor is located in the center of the dashboard above the radio.

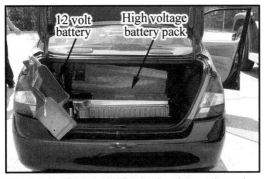

The trunk of this hybrid contains the two batteries, a high-voltage battery and a twelve-volt battery.

The high-voltage cables are isolated from the chassis in this channel beneath the floor pan.

Fig. 3–9 Hybrid vehicles—no longer the car of the future

Fire control and extinguishment

Post-crash engine compartment fires, while not common, do occur on occasion. Should a fire erupt in a vehicle containing trapped victims, aggressive fire suppression activities are needed. If fire erupts in the engine compartment after a crash, the hood will probably have significant damage, making access and extinguishment easy. A 1-3/4 in. hose line should be pulled and used for a quick and powerful knockdown; booster lines with their low volume flow and poor streams should be left on the apparatus. For quick deployment of a hose line, many crews use 100-ft, bumper-mounted lines for car fires and extrications. These lines are just the right length, provide plenty of water, and are easy to reload.

Fig. 3–10 Front bumper line outfitted with 100 ft of 1¾ in. hose.

When responsible for extinguishing a fire in the engine compartment, the crew member on the nozzle should remember a few tips.

1. SCBA should be worn. SCBAs not only protect the crew member on the nozzle, but they allow work to be done in heavy smoke conditions. When a person is trapped in a burning vehicle, the firefighter on the nozzle should be able to work without *any* distractions or annoyances caused by heat or smoke.

2. Pressurized auto parts can fail when heated. Tires, bumper systems, and hood lifters can all rupture when heated and should be avoided.

3. If there's no reason to stand directly in front of the burning area, don't. If the fire is blowing out from under the fender on one side of the vehicle, it may be best if the hood is forced open and the fire extinguished from the other side of the vehicle.

In the event an engine compartment fire starts and the hood is fully intact and not bent, crew members will have to open it up to extinguish the fire. If conditions are appropriate, attempts to locate and use the hood release in the passenger compartment can be made. If there is a significant fire under the hood, more aggressive procedures should be used. There are many methods for opening the hood in a hurry, but the most effective, time-honored method is to use the adz and the claw of a Halligan to pry up the side of the hood. Following are the two big advantages to using the Halligan:

- There's never a problem with a Halligan not starting or running out of fuel.

- Its operational success rate is nearly 100%.

Even though the basic Halligan procedure is the favorite of most firefighters, it's good to have other options in case the Halligan is being used to help rescue the victim or is otherwise unavailable. A saw outfitted with a quality aluminum-oxide blade can be used to open up a hood, but the process is slower than using a Halligan. While the saw is slower than the Halligan when making an initial attack, it can be used to good advantage when the engine compartment needs to be fully exposed. Full exposure may be needed to perform overhaul or to access all areas of the engine compartment to extinguish the fire.

Like extrication procedures, opening up a hood with forcible entry tools can pose some hazards. When using the tools, crew members should remember that radiators are hot, pressurized containers that can spray hot coolant when cut open. If punctured, batteries can pose a problem, as can the lifters used to raise a hood. When selecting a method to open a hood, all of these factors should be taken into consideration. While the problems created by forcing a hood open are usually small, there's no sense in looking for trouble. The following procedures include the simple Halligan method and several other methods that can be tried in the junk yard and compared for speed and effectiveness.

Vehicle stabilization—stopping large movements

Working on a vehicle that's susceptible to sudden, uncontrolled movement is dangerous to both the trapped occupants and the crew members performing the extrication. Large movements can occur if a vehicle is on a steep embankment or slipping into a body of water.

To prevent any additional movement, a come-along, winch, or rope can be used to hold a vehicle in position. After being secured, if necessary, the vehicle can be pulled into a position that allows the occupant to be extricated.

Come-alongs can be used for small operations, but wreckers or apparatus equipped with winches should be used for bigger operations. When a wrecker is needed, the officer should be clear in the request that the wrecker is needed for a rescue operation and not a routine towing operation. If working on a busy highway or expressway while waiting for the requested wrecker, the officer should watch for any passing wrecker that could be flagged down and used in the rescue operation.

Vehicle stabilization—stopping small movements

Rollover crashes can result in a vehicle coming to rest on its roof or side, creating a situation where the victims may be suspended by their seat belts or

Fig. 3–11 After leaving the road, this vehicle came to rest in a canal with its rear wheels dangerously close to the edge of the embankment. Canals in this area can be very deep, posing a serious threat of drowning to not only the occupant but also the crew involved in the victim removal.

Fig. 3–12 Initially a come-along and ratchet strap were used to stabilize the vehicle. The come-along and ratchet strap were both attached to the guard rail. When a towing company wrecker arrived on the scene, a winch was connected to the vehicle for additional holding power.

Stabilization

in a pile near the ground. Access to the victims in these types of crashes is different than when the vehicle is on all four tires because the doors are usually damaged or inaccessible. Procedures needed to access and free the victims can cause the vehicle to rock, collapse, or roll over, creating additional risk to victims and crew alike. Vehicles that have been involved in severe crashes but remained on all four tires can also have problems caused by structural weakening, requiring support to prevent additional structural failure.

To resolve these problems and to create a safe work environment, vehicles involved in crashes that leave them in an unstable condition need to stabilized. The degree of stabilization performed should be determined after the officer performs a size-up and risk/benefit analysis of tactical options. In complex and lengthy operations, time spent in the beginning to create a stable work environment often proves to be a benefit later in the operation. By stabilizing a vehicle that's resting on its side, crews can be more aggressive in their actions when working on the vehicle because there will be less concern with the vehicle rolling over. The same holds true for vehicles resting on their roofs or in some other unusual position.

The procedures used to stabilize the vehicle are determined by the following factors:

- How the vehicle may move during the operation as the metal of the vehicle is moved or cut away
- Access to the vehicle and the equipment needed to stabilize it
- The amount of unobstructed working area needed by the crew to perform the extrication

Ultimately, the procedure selected for stabilizing a vehicle will be based in great part on the tools available to the crew. Fortunately, cribbing is carried on many types of fire apparatus and is often the best and simplest solution to stability problems (see Figs 2–1, 2–2).

Figs. 3–13 & 3–14 These two vehicles required the same type of procedures to extricate the victims. The difference between the generic cribbing and specialized stabilization equipment is in the fit and stability. The standard long 4x4 has been used successfully for years and will continue to be used in the future, but lacks a built-in method of securing the 4x4 to keep it from slipping out. New types of equipment however, provide for quick and more secure operations.

A few pieces of 4x4 cribbing, some wedges, or some rope may be all that is needed to create a reasonably safe work area at a crash scene. Other types of stabilization specialty tools can be helpful and should be used when needed and available.

Whichever methods and tools of stabilization are selected, the officer and crew should be able to adapt to the situation at hand and modify the basic procedures to fit their needs. Intuition, training, and the ability to adapt are important traits in being able to evaluate a stabilization problem and solve it with available resources. The procedures that follow in the Procedures section can be used as a foundation, suitable for adaptation in any given situation.

Stabilization of the Victim

The procedures used to evaluate and treat trapped victims will vary from one department to another in response to local capabilities. Local protocol will determine the type of care provided to the victim during the extrication.

The key is that evaluation and treatment be performed during the extrication (see Figs. 3–15 & 3–16). To provide care, emergency medical service (EMS) providers usually assign one crew member in bunker gear to the inside of the vehicle to perform the following actions:

- Assess the victim's injuries

- Reassure and inform the victim of the situation and the basic action being taken to remove them from the vehicle

- Protect the victim from additional injury by shielding the victim from glass as it's removed and monitoring the movement of metal

- Notify the officer in charge of the extrication of any changes in the victim's condition that may require a change in the methods being used to extricate the victim

- Start medical care of the victim including, but not limited to, the application of oxygen, control of bleeding, spinal immobilization and packaging. During longer operations, intravenous fluids (IVs) can be established along with other advanced life support treatment.

Once the victim has been disentangled from the wreckage, the crew responsible for operating the tools can back out and let the crew responsible for the victim's medical care move in. When the extrication crew is backed out, it gives them a chance to rest for a minute and prepare to resolve any problems that pop up during the removal of the victim from the wreckage. However, there will be times when both the extrication crew and medical crew will have to work together to remove a heavy or badly injured victim. The division of responsibilities is not meant to discourage helping other crews, but instead to help identify and clarify roles and responsibilities at the crash scene.

Stabilization

Figs. 3–15 & 3–16 This tractor trailer underride crash resulted in entrapment with very limited access to the driver resulting in only her arm being visible from beneath the trailer.

victims. Tool operations, coupled with the simultaneous medical treatment of trapped victims, reduce the amount of on-scene time needed for medical treatment after the victims have been extricated. Strong scene and vehicle control results in smooth victim treatment and timely transportation to the hospital or trauma center.

Summary

Crews owe it to themselves and crash victims to get crash scenes under control as quickly as possible. Good scene management results in less danger to everyone and allows the extrication and medical crews to work confidently and aggressively to free trapped victims without worrying about their own safety. Vehicles that are stable pose less risk of undesirable movement, offering more tactical options that can be utilized to free trapped

Procedures

Vehicle Extrication: A Practical Guide

Procedure 1
Engine Compartment Access Using a Halligan Adz

When extinguishing an engine compartment fire, prying the edge of the hood up provides the simplest, fastest, and most reliable option. The first step, after readying the hose line, is to insert the adz between the hood and the fender. If there is a tight fit, the Halligan may need to be struck with a sledge hammer or ax. With the adz inserted an inch or so, the handle of the Halligan is swung around, spreading the fender and hood apart.

The Halligan is then removed and positioned so the adz can be used to pry the hood upward. To get a good bite with the adz, it may be necessary to position the Halligan almost flat against the hood before prying up.

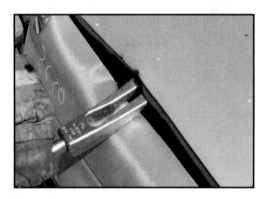

For additional height, the adz can be removed, the Halligan swung around, and the claw end inserted for prying. If, for some reason, the opening must be closed again, the Halligan, sledge, or ax can be used to make downward blows on the hood, flattening it.

| Stabilization |

Procedure 2
Engine Compartment Access Using a Halligan Pike to Pry a Corner

Prying up the corner of the hood with a Halligan will provide access to the engine compartment but has drawbacks that make it more hazardous than other procedures. To lift the corner of the hood, the pike of the Halligan is positioned about a foot from both the front and side edges of the hood. The Halligan is then struck with a sledge hammer or ax, driving the pike through the sheet metal of the hood. The drawback to this procedure is the likelihood of striking a battery with the Halligan. In the case of diesel engines with dual batteries, the likelihood increases to almost 100%.

Once the Halligan has penetrated the hood, the handle of the Halligan is pushed toward the center of the hood. By pushing toward the center of the hood, a crease will develop in the metal, allowing it to fold.

Once the corner is pried up, the claw of the Halligan can be used along the edge of the hood to pry it up further if necessary.

Vehicle Extrication: A Practical Guide

Procedure 3
Engine Compartment Access Using a Halligan Pike to Punch a Hole

In cases when only a dry chemical extinguisher is available for fire suppression, it may be best to limit the amount of fresh air introduced into an engine compartment that is burning. If the side of the hood was pried up and the extinguisher was unable to extinguish the fire, crewmembers would have to try to bend the hood down again to keep the fire from growing. To avoid this problem, a smaller opening can be made with the pike of the Halligan. The pike is positioned about a foot inward from the outer edge of the hood and is struck with an ax or sledge hammer.

To ensure a large enough hole for an extinguisher nozzle, the Halligan is struck until the pike completely penetrates the hood. By burying the pike in the hood, there is a greater chance of penetrating both layers of the hood assembly.

The handle is then swung 180 degrees away from the crewmember with the tool. The crewmember will be over the gap between the hood and the fender and must be aware of any fire that extends through the gap.

| Stabilization |

Procedure 3 *(continued)*
Engine Compartment Access Using a Halligan Pike to Punch a Hole

The handle is then pulled upward. This action increases the size of the hole, making it easier to insert an extinguisher nozzle. Additionally, the prying action will create an opening toward the center part of the engine compartment.

The head of the Halligan serves as the fulcrum for the prying action.

Once a large hole has been created, the nozzle of a dry chemical extinguisher can be inserted and discharged.

Vehicle Extrication: A Practical Guide

Procedure 4
Engine Compartment Access Using a Saw (Front Latch Cuts)

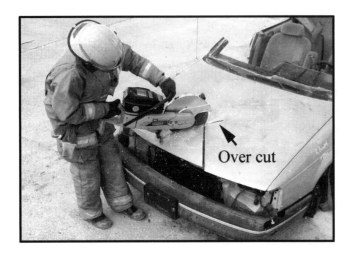

Cutting the hood to defeat the front latch is probably the easiest cut to make with the saw. By eliminating the latch, the hood can be lifted on the hinges and lifters. When considering this procedure, it should be confirmed that the vehicle has only one latch (not two) and that it is in the front and not in the rear of the hood. Overlapping cuts are made in the shape of an inverted V.

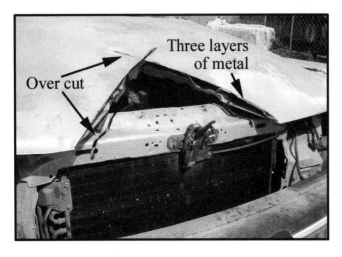

The drawback to this procedure is the possibility of cutting into a hot pressurized radiator. Additionally, if the bumper system has been exposed to fire, there is a chance of over-pressurization and failure. If this procedure is used, the cuts should be longer and deeper than anticipated. This will ensure that the cuts are complete and connected.

With the latch removed from the hood assembly, the hood can be raised.

Stabilization

Procedure 5
Engine Compartment Access Using a Saw (Center Cuts)

For good access to most areas of the engine compartment, the center of the hood can be cut and opened up. An X is cut in the center of the hood, making sure that the cuts are deep enough to penetrate multiple layers of metal.

With the X cut, the adz of the Halligan can be used to hook the wedges of metal and pull them up.

The importance of cutting deep can be seen in this photograph. If the hood bracing is not cut completely, it may be difficult to pry the sheet metal up. The insulation seen to the left of the brace can be pulled up or pushed down with the Halligan.

Procedure 6
Engine Compartment Access Using a Saw (Straight Cuts)

This procedure results in the hood being cut into two pieces. Working from the side of the vehicle, the saw operator reaches across with the saw to just beyond the halfway point of the hood. Once in position, the saw is brought up to a speed high enough that the blade will continue to spin when it is lowered into the hood. Once the blade has penetrated the metal, the saw is brought up to full speed. The saw is then pulled back in the direction of the operator. To avoid not cutting far enough, the cut should extend well into the fender.

With the first cut completed, a similar cut is made from the opposite side of the vehicle. To avoid any problem with uncompleted cuts, the second cut should overlap the first cut's starting point.

With the front half of the hood cut away from the back half, the overall weight of the hood is cut in half. This means the hood lifters or springs will push the hood up as the cut is being completed. The saw operator must be prepared for blade binding and sudden release and elevation of the rear half of the hood. To minimize the chance of injury, the arm that holds the front of the saw should be the arm closest to the front of the vehicle.

Stabilization

Procedure 7
Engine Compartment Access Using a Saw (Corner Cuts)

Most automobiles have hinges at both rear corners of the hood to assist in raising the hood. If cuts are made to disconnect these hinges from the rest of the hood, the hood can be pivoted upward at the front latch. To make the cuts, the saw operator must visualize where the ends of the hinges are and cut in front of them. The cuts extend from the rear of the hood to the fender. To avoid missing parts of the hood structure, all cuts should be made a little longer and deeper than anticipated.

A couple of drawbacks to this procedure is the possibility of cutting into the strut tower and cutting through the gas operated lifters. Additionally, this procedure requires two cuts and movement from one side of the vehicle to the other.

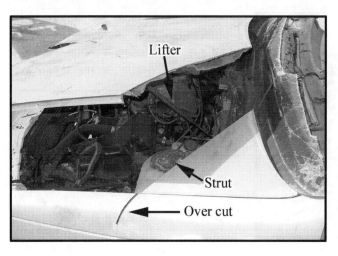

In this case, the hinges were cut away and the lifters stayed intact, pushing the hood upward. In other cases, the hinges may be connected to springs that will cause the corners to spring up as soon as the corner is completely severed by the saw. The saw operator should be aware of this possibility and be in a position at the end of the cut to avoid being struck by a corner.

Procedure 8
Using Cribbing to Stabilize a Vehicle on Its Side

When cribbing a vehicle to prevent it from rolling onto its roof, base the location of the cribbing on both the stability of the vehicle and other extrication procedures that will be performed. Cribbing at the base of the "A" post is usually a good choice but does not provide as wide a base as cribbing that is applied out at the drip edge. When placing cribbing at the drip edge, the top of the roof posts provide good strength. If the roof is to be removed, cribbing at the drip edge would serve no purpose once the roof is removed.

Wheels and tires provide a wide base, making rollover less likely. If only wedges are available, two or three wedges go a long way in preventing a rollover.

If wedges are used, there may be a gap between the body of the vehicle and the wedges. If a long 4x4 is available, placing it between the vehicle and the wedges will firm up the cribbing.

Stabilization

Procedure 8 *(continued)*
Using Cribbing to Stabilize a Vehicle on Its Side

When this vehicle came to rest, it settled on the wheels and the main body of the vehicle but not the roof. After settling in this position, the drip edge of the roof was a considerable distance off the ground, requiring taller cribbing.

To match the different angles of the roof, box cribbing can be built with a combination of 2x4s, 4x4s, and wedges. The wedges in the foreground are positioned for greatest strength by the positioning of the 2x4 directly beneath them. The load is then transferred to the ground without any gaps. The box crib in the background is not as strong because there is a gap under the wedge instead of another piece of cribbing.

The height of the cribbing in the foreground worked out well for the positioning of the wedge to snug up the entire box crib. Toward the center part of the roof, an additional piece of 2x4 was needed to fill the gap between the 4x4 and the wedge. For best results, the crewmember responsible for the cribbing should be resourceful in making the cribbing fit the space.

Procedure 9
Stabilizing a Vehicle to Prevent Further Forward Movement

This scenario depicts a vehicle that has left the roadway and has come to rest with the front wheels suspended over a culvert. The edge of the concrete wall has become a fulcrum point, causing the vehicle to rock up and down. The fulcrum point is so close to the center that the rear of the vehicle can be lifted with one finger. The key point here is that the smallest amount of downward force in the front of the vehicle could cause the vehicle to slip forward off the wall.

To prevent this action, a 3-ton come-along is used to hold the vehicle in position. A winch would provide a more powerful solution and should be used if available. After a good size-up, the officer may direct crewmembers to physically hold the vehicle to prevent the vehicle from slipping while the system is built. The first step is to attach a chain to a sturdy structural member at the rear of the vehicle.

The come-along is attached to the master link. The crewmember rigging the come-along should try to continually observe the vehicle in the event that it slips off the wall, pulling the equipment out of the operator's hands.

Stabilization

Procedure 9 *(continued)*
Stabilizing a Vehicle to Prevent Further Forward Movement

The come-along is walked back to the anchor chain. The come-along is kept in a 2:1 mechanical advantage configuration for maximum pulling power. If the distance between the master links is too long for the 2:1 configuration, the cable hook can be used instead of the block.

The come-along is connected to the master link of the anchor chain. Any excess chain is removed with the chain shortener on the anchor chain.

The officer calls for the slack to be taken out of the system, then has the chain hooks checked to ensure the rigging is correct after removing the slack.

Vehicle Extrication: A Practical Guide

Procedure 9 *(continued)*
Stabilizing a Vehicle to Prevent Further Forward Movement

The telescoping handle is estended and locked in position for maximum leverage. The officer orders the come-along to be operated.

The length of the telescoping handle makes it difficult to operate in a purely vertical plane, so the operator can rotate it down to the side for easier operation.

The officer should decide if the vehicle should be simply secured until the arrival of a vehicle with a winch, or if there is more to be gained by pulling the vehicle completely out of the location.

Stabilization

Procedure 10
Stabilizing a Vehicle on Its Side with a Come-along

Considerable stability can be gained using a basic come-along set and a few pieces of cribbing when dealing with a vehicle on its side. After cribbing is positioned, one chain is connected to an anchor point and the other chain is connected to a high, sturdy point on the vehicle. If the distance between the anchor and the vehicle is short enough, the come-along can be used in the 2:1 mechanical advantage mode.

Most come-along sets consist of a come-along and two chains, 12–14 ft in length. If three chains are available, one chain can be set to the anchor and the other two can be attached to the vehicle creating a more stable situation. After the chains are attached to the vehicle, they are brought back to the master link. If one chain is longer than the other, the chain shortener should be used to take up the slack chain. This will help in creating an even pull.

Instead of a sling hook at the end of the chain, this chain set has a grab hook. The chain was wrapped around the front end assembly then hooked back to the chain.

Vehicle Extrication: A Practical Guide

Procedure 10 *(continued)*
Stabilizing a Vehicle on Its Side with a Come-along

With both chains attached high on the vehicle, the master links are brought back to the come-along block. In the photograph, the chains were about the same length when connected to the block, making use of the chain shorteners unnecessary.

The best anchor positioning results in the line of the cable being perpendicular to the vehicle. Circumstances will control the choice of anchors, but, if given a choice, use fire apparatus or another heavy vehicle. The officer in charge of the vehicle must take steps to ensure that the vehicle is not driven away by somebody unaware that it is being used as an anchor.

With the system built, the officer directs a crewmember to take the slack out of the come-along. When the slack cable is taken up, the hooks and chains are checked to ensure they are in the proper position. With close observation of the vehicle, the officer or designee directs the come-along operator to operate the come-along until enough tension is on the cable to prevent the vehicle from rolling over.

Stabilization

Procedure 11
Using a Long 4x4 to Stabilize a Vehicle on Its Side

A simple approach to stabilizing a vehicle on its side utilizes a couple of 4x4s butted between a sturdy part of the chassis and the ground. While the choices may be limited, the contact point on the chassis should be as high as possible to reduce the chance of the four by four kicking out when under load. Dirt or sand provides a better base than asphalt or concrete in this situation.

To reduce the chance of the 4x4 kicking out, a ratchet strap can be wrapped around the 4x4 and then attached to the chassis. After the cargo strap is put under tension, the 4x4 is less likely to slip along the asphalt or concrete.

If there are no ratchet straps available, webbing or strapping can be used to secure the 4x4. To take the slack out of the strap, a pair of opposing wedges is placed between the 4x4 and the strap. With the strap snug, the wedges are simultaneously driven toward one another. As the wedges slide past one another, the height of the pair increases, removing more slack from the strap. By reducing the slack from the strap loop, the long 4x4 becomes more stable and less likely to shift.

Vehicle Extrication: A Practical Guide

Procedure 12
Stabilizing a Vehicle from Both Sides Using a Capabear Stabilization Device

The best procedure for stabilizing a vehicle on its side involves the use of cribbing and some type of long support. This vehicle has been supported on the roof side with the use of wedges, large cribbing, step chocks and a Capabear. These devices bridge the gaps between the ground and the passenger side "A" post, the driver's side hinge pillar, the top of the "B" post and the top of the "C" post.

The head of the Capabear serves as a saddle for the "A" post.

The bottom of the Capabear needs to be secured, and in this case the option of running it to the opposite side of the vehicle is shown. Before positioning a strap over damaged sheet metal, the metal should be checked for sharp edges that can easily cut through a strap under tension.

Stabilization

Procedure 12 *(continued)*
Stabilizing a Vehicle from Both Sides Using a Capabear Stabilization Device

Another option is to wrap the strap around the "A" post and bring it back to the ratchet. Again, any sharp remnants of metal and glass will cut a strap, requiring a quick evaluation before being used.

The chassis side of this vehicle first received 3 large wood wedges, then 2 additional Capabears. The 4x4s used in the photograph were of a nominal length, as would be the case for most apparatus carrying long 4x4s. To achieve a stronger configuration, the supports could be positioned a little more vertically. This would require the 4x4s to be shortened with a chain saw or circular saw.

Careful placement of the upper end of the support is necessary. The support in the foreground has a very small contact area with the vehicle, making it more likely to be accidentally knocked out while the support in the background is making much better contact.

Procedure 13
Stabilizing a Vehicle from One Side

If a vehicle has rolled over and come to rest on its side, it is usually best to stabilize from both sides of the vehicle for maximum stability. There may be situations, however, when the vehicle must be stabilized from one side only. This could occur because of difficult terrain, limited access, severe damage, or the need to have a clear working area for the crewmembers performing the extrication. In the situation depicted in the photos, the roof side of the vehicle will be considered unusable for stabilization, requiring all the work to be done from the chassis side of the vehicle.

To get started, wood wedges and other cribbing are put in place on both sides of the vehicle if possible, to provide a minimum amount of support.

If it appears that the vehicle is at great risk of rolling over onto its roof, it should be tied back to prevent it from rolling before other steps are undertaken. For best support and leverage, the chains should be attached to the chassis at high spots.

Stabilization

Procedure 13 *(continued)*
Stabilizing a Vehicle from One Side

The anchor (in this case fire apparatus) should be large and heavy enough that it won't slide forward if a load were suddenly put on the system. In this procedure, a come-along was used to secure the vehicle, but the stronger bumper-mounted winch would be another option. When a vehicle is resting on its side, not much force is needed to unintentionally pull the vehicle over. This can be a problem when using a winch because the operator won't have a feel for the load that is present when using a come-along.

For good support, the ground pads are attached to this Paratech brand strut system. The angle is set to provide good strength and little chance of sliding out of position.

The strut is telescoped out to make good contact with a strong, large point on the chassis. Once in position, the collar is spun down to lock the strut to the desired length.

Procedure 13 *(continued)*
Stabilizing a Vehicle from One Side

With the strut in position, the base is given support by the use of two straps per strut. In this system, instead of one strap being positioned in line with the support, the straps extend to the sides of the strut.

Suitable anchor points may be difficult, as is the case in this photograph. This structural member is used for transporting a vehicle in the normal position. The load would be straight down instead of toward the front, as is the case in this photograph. When selecting anchor points, consideration should be given to the load and the suitability of the anchor. Some anchors, while not preferred, may have to be used to complete the extrication.

With the hook attached to the anchor, the slack is taken out of the strap. To avoid lateral shifting of the strut, final tensioning should be done after the second strap is in place.

Stabilization

Procedure 13 *(continued)*
Stabilizing a Vehicle from One Side

When the second strap has been attached to the chassis, the slack is taken out, and both straps are adjusted to the desired tension.

A second strut is installed for additional stability. The strap procedures are the same.

If a tieback was not done at the beginning of the procedure, it can be done after the struts are in place. With the struts in place and a come-along or winch attached to the vehicle, final adjustments can be made. If a winch is used, the operator must be careful to use only enough power to secure the vehicle in position and not exert so much force that the vehicle starts to be pulled over the struts.

Procedure 14
Stabilizing a Vehicle on Its Roof with Step Chocks, Capabears, and Box Cribbing

To keep a vehicle from rocking on its roof, step chocks can be placed under the roof to fill the area between the roof and the ground. It is important to note that this provides no support to the main portion of the vehicle's body. Additionally, if the doors are opened, the vertical support provided by the window frames would be removed. This procedure is used routinely by crews in the field primarily because their options are limited by the tools they carry on the rigs. The officer should carefully examine the post-crash integrity of the vehicle before assuming this procedure is the best way to stabilize the vehicle.

Another option is the use of short Capabears. To be used successfully, the top of the supports must be set against a solid portion of the vehicle. The vehicle in the photograph had a sturdy bumper that was suitable for being used as a structural member. Many bumpers are not sturdy enough to support loads and should not be used for stabilization.

The best choice is to build box cribs from the ground to the body of the vehicle. This cribbing configuration provides good support and can easily be assembled when the surface of the vehicle is parallel to the ground. If not parallel with the ground, the procedure becomes a little more difficult, requiring more planning and the use of wedges.

Procedure 15
Stabilization on the Roof

If there is any question about the integrity of the roof posts or if the overturned vehicle is a convertible, cribbing should be placed between the ground and the main body of the vehicle. Should the roof posts fail without warning or as a result of a door being opened, the cribbing will help prevent the vehicle from collapsing to the ground.

Box cribbing takes up a considerable amount of space so crewmembers performing the stabilization job should anticipate how and where other crewmembers and equipment will enter the vehicle, and from where the victim will be removed after being placed on a backboard. If the entry is to be through the rear windshield, two sets of box cribs will make for a tight fit.

If building a box crib at one spot does not work out, other locations can be considered after estimating their overall strength.

Vehicle Extrication: A Practical Guide

Procedure 16
Stabilizing an Overturned Vehicle with Stump Screw Jacks

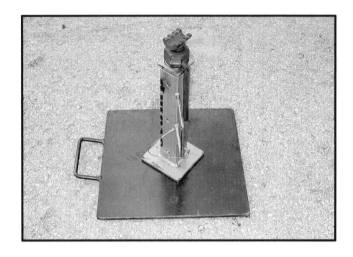

The Stump screw jacks can provide quick stabilization for overturned vehicles and can be especially useful when a vehicle's roof posts are badly damaged, increasing the risk of collapse. To use the jack, it is first attached to a large flat pad similar to an outrigger pad.

After assembling the jack, slide it into the desired position directly beneath a solid and structurally sound part of the vehicle. Speaker holes in the rear deck should be avoided; the intersection of the trunk lid and the body of the vehicle would be a better contact point. Once in position, the adjustment pins are removed and the upper part of the jack is telescoped out to close the gap between the jack and the vehicle.

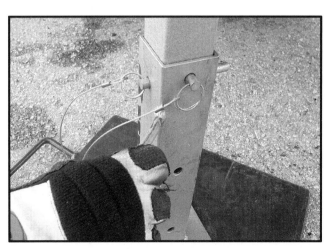

With the upper section of the jack extended, the adjustment pins are reinserted to lock the jack at the appropriate height. After the jack height is set, there will be a small gap between the jack head and the vehicle.

Stabilization

Procedure 16 *(continued)*
Stabilizing an Overturned Vehicle with Stump Screw Jacks

The gap between the vehicle and the jack is filled by extending the screw and head. To extend the screw, the handle is used to spin the operating nut.

A second jack will provide uniform support and a third point of contact with the ground.

With two Stump jacks in place, the danger associated with collapsing roof posts is reduced. Additionally, the pressure of the screw jack will help keep the jack in place if it were to be bumped. With their large foot print and small size, the Stump jacks provide crewmembers with more room to work in the rear window area than if box cribbing were used.

Vehicle Extrication: A Practical Guide

Procedure 17
Stabilizing a Vehicle on a Jersey Barrier

This vehicle is in a fairly stable position with three points of contact with the ground: two points are made with the front tires and the third is made via the concrete highway barrier wall. These types of barriers are very common in construction areas on limited-access highways. While this vehicle is stable, a real crash may leave the vehicle in a position not so stable.

To minimize the side-to-side movement of the vehicle that would occur when treating and removing victims, two short Capabears have been put in place to provide two additional points of contact. A ratchet strap connects the two bases.

To prevent the vehicle from slipping forward, a come-along is used to tie the vehicle back. The block was attached to an eye used to secure the vehicle when it is being transported. This provided a good attachment point, eliminating the need for a chain. In this type of construction area, there usually won't be any good anchor points other than vehicles or heavy equipment that is driven to the scene. To create an anchor, a pry bar was inserted in the hole in the barrier. The chain is then shortened to the correct length. Operating the come-along may only take a light touch to prevent the vehicle from being pulled further than intended.

4

Door and Side Procedures

TERMINOLOGY OF COMPONENTS AND PROCEDURES

When trying to communicate a plan, it's nice when everybody involved in the plan is using the same terminology. When discussing the side of an automobile, the common terms seem pretty basic: *door*, *hinge*, and *Nader bolt*. Other terms are a little fuzzier and can cause some problems when trying to communicate: *pillars*, *posts*, *latches*, and *locks*. This isn't surprising, considering that even automobile manufacturers use slightly different terms for similar components.

In this book, the left side, street side, and driver's side are considered the same thing and are referred to as the *driver's side*. The other side of the vehicle is referred to as the *passenger's side*, with front and rear used to further identify the specific locations. The terms *driver's* and *passenger's sides* are preferred to left and right because they cause less confusion when the vehicle is on its side or upside down.

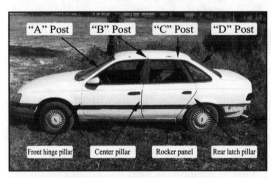

Fig. 4–1 Vehicle components terminology.

The roof posts are identified by letter; the "A" post is at the front, the "B" post is usually near the center, the "C" and possibly "D" are at the rear. The roof posts extend downward about 20 in.

from the roof to the bottom of the window opening. At that point, they become pillars that support latches and hinges. While it may seem like splitting hairs, misunderstandings about whether a post or pillar is supposed to be cut can cause problems. Even with pillars, the terminology can vary a little. In a two-door vehicle, the hinges are connected to the front-hinge pillar, forward-hinge pillar, or hinge pillar; these are different terms for the same thing. The latch striker (Nader bolt) is attached to the latch pillar. In a four-door vehicle, the center pillar is most easily described as the *center pillar*, but may also be described as the *rear-hinge pillar*. In the interest of simplicity and clarity, the terms *front-hinge pillar*, *center pillar*, and *latch* (or *rear-latch pillar*) will be used to describe those specific components.

Fig. 4–2 Vehicle components terminology.

The *rocker panel* is the part of the body that faces the bottom surface of the door when the door is closed. The *fender* is the piece of sheet metal that covers the wheel area. The *latch striker* is more commonly know as the *Nader bolt*—a term adopted in recognition of Ralph Nader's contributions to automobile safety.

Procedure names are usually self explanatory but can cause a little confusion if popping a door open is described as "forcible separation of door latch components with an engine-powered hydraulic spreader tool." *Popping a door* doesn't sound very sophisticated, but it's the term most commonly used around the firehouse and at the crash scene, making it the best term to use during training. Terms used to describe other door actions include the term *widening*—a procedure used to force the door open wider than the hinges would normally allow and the term *removal*, which is self-explanatory. The procedure of removing the front and rear doors, the "B" post, and center pillar in one big piece has several names including *rip and blitz*, *barn door*, and *side-out*. The term *side-out* will be used here.

THE SIZE-UP AND ACTION PLAN

Popping a door open with a hydraulic spreader tool is the most common extrication procedure performed by firefighters and is often the only procedure needed to extricate a victim. Usually the door latch mechanism is damaged or the sheet metal is bent in such a way that causes it to bind, preventing the door from swinging freely. When performing the size-up, the officer and crew should quickly try to determine where and how the door is stuck before developing a plan for action. The size-up may reveal that the door is simply locked, or that one of the handles is damaged. If this appears to be the case, the door can be unlocked and both handles are tried simultaneously. On occasion, and with a stroke of good luck, the door will pop open, eliminating the need for the big tools.

Door and Side Procedures

If the door is badly damaged and it's apparent the victim will need to be extricated, the officer should remember to answer the simple, but most important question: *What's keeping the victim from coming out of the vehicle?* The answer to the question may be surprising. While the door may initially appear to be the sole problem, a closer examination may reveal that after the door is opened, the "B" post and center pillar may continue to block the victim's removal. This type of entrapment occurs in side-impact crashes when the center pillar is forced against the victim's seat back, pushing it and the upper body of the victim into the rear seat area. Early recognition of all of the problems can lead to a different set of solutions that, if not chosen, can result in a big waste of time and effort. This loss of time, while frustrating for the crew members, could have a serious impact on the victim's chance of survival.

After determining that the center pillar is going to be a problem, the officer can consider the four basic tactical options and determine which option, or combination of options, would work best. In the case of the victim trapped behind the "B" post, the first option (spreading) may or may not work, depending on the damage configuration. To spread, there needs to be a push-off point for the spreaders or ram. A ram set up to push from one center pillar to the other may push the problem pillar away, but that may not free the victim.

The second tactical option (pulling) could work if the doors were removed, the "B" post cut and relief cuts made at the bottom of the center pillar. If all of these actions are possible, a pulling tool and an anchor point would then be needed to pull the "B" post. These steps, while workable, intuitively don't seem to be the right answer.

The third tactical option (removing) is probably the best option. In this situation, the center pillar and both doors can be removed by performing the side-out procedure.

The fourth tactical option (securing in place) wouldn't help at all. At best, the center pillar could be bypassed if the victim could be slid out of the vehicle onto a long backboard through a rear hatchback window. Again, after identifying the cause of entrapment and considering all the options, the removal option appears to be the best choice.

In another crash, the "B" post may not be an issue. Adequate access may be possible if the front door is simply moved out of the way. After a door is popped open with the spreaders, it may not swing open because of damage to the hinge area. Again, the four tactical options are considered, with three of the four being suitable for opening up the space. The door can be spread open by placing the tips of the spreader between the door and the front-hinge pillar and opening the tips. The pulling option can be utilized by pulling the door open with a come-along. The third option (removing) can be utilized with hydraulic cutters, spreaders, or an air chisel. The fourth tactical option (securing) won't help in this situation.

The comparison of these two similar crash configurations illustrates the importance of a good size-up before developing an action plan. Only by performing a size-up, then considering the tactical options, can an action plan be put together.

Selecting the Best Procedures and Tools

With a good size-up completed, the officer and crew can select the extrication procedures. The factor with the greatest impact on procedure selection is the equipment available to the crew. If the apparatus that the crew is riding on only has hydraulic tools, the choice is simple. If the tasks can be completed with those tools, a plan can be developed and put into action. If the officer sees that other tools will be needed for the extrication, the request for other units should be made promptly to minimize any delays in extricating the victim.

The skills and abilities of the crew also have a big impact on the selection of tactics. The procedure of creating a third door on a two-door vehicle can be completed with a set of hydraulic tools, but a far more effective tool choice is the air chisel. When hydraulic tools are used to create the third door, only a few simple cuts are needed before brute force is used to pry the vehicle open. While the procedure used for creating the third door with hydraulic tools is simple, the end result often isn't always what's needed. On the other hand, a quality air chisel in the hands of a skilled operator can open up a large hole for the removal of a backseat victim in less than three minutes. When given a choice among tool systems, the officer should consider the skill of the operators, the quality of the tools, and the likelihood of timely completion of the procedure before selecting a plan of action.

If the officer has the good fortune of having several tools available, multiple tools can be put to work simultaneously rather than just one tool. The crew members will have to pay attention and stay out of each other's way, but a well-trained crew should be able to operate in this manner without a problem. The officer can keep an eye on the crew members doing the work and help coordinate their efforts.

The Procedure Steps

The procedure steps described focus on the procedure itself and do not include the universal steps of stabilizing the scene, vehicle, and victim. Information about hazard assessment is addressed in chapter 5 and should be applied as appropriate to the procedures that follow. Additionally, the procedures for removing tempered glass and creating a purchase point are described at the beginning of the section and should be used as needed. Generally, tempered glass should be removed any time an extrication procedure will apply force against the glass, directly or indirectly,

Fig. 4–3 As this door was spread open and the inner and outer metal separated, the glass broke in an uncontrolled manner. The position of the glass increases the chance of glass pieces being propelled up and under face shields.

Door and Side Procedures

even when the window is rolled down. Tempered glass is strong and flexible—two characteristics that allow it to be placed under considerable load before breaking. This can be a problem when a door is being forced open and the sheet metal on the inside and outside of the door separates.

The glass that is under a load can break without warning, propelling glass fragments over a considerable distance. It's best to break the glass under controlled conditions.

The procedures that follow are demonstrated with tools that are known to give good results. There may be variations of the procedures that work just as well, or perhaps better than the basic procedures demonstrated here. If a superior procedure is known, the officer and crew should utilize it as needed.

PROCEDURES

Vehicle Extrication: A Practical Guide

Procedure 1
Removing Tempered Glass with a Center Punch

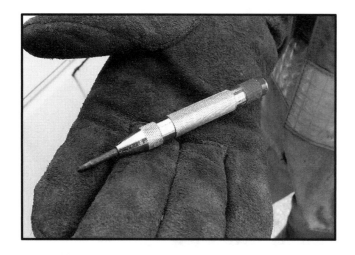

The spring-loaded center punch is a common tool used for removing tempered glass. The internal mechanism of the tool allows force to be applied to the rear of the tool until a specific point is reached, at which time the force is transferred to the punch part of the tool. To use the spring-loaded center punch, it is positioned against the glass, then steady force is applied to the rear of the punch until it releases and breaks the glass.

If the point on the punch is dull or rounded, several attempts may be needed to obtain the desired result. Internal parts of the punch are prone to rusting if they become wet. To reduce the chance of rust and malfunction, the punch should be disassembled, cleaned and lubricated after getting wet.

Even after the glass is broken, tempered glass may stay in place. If possible, the glass should be pulled to the outside away from the occupants. A windshield saw or other tool may be useful in pulling the glass away from the frame.

Door and Side Procedures

Procedure 1 *(continued)*
Removing Tempered Glass with a Center Punch

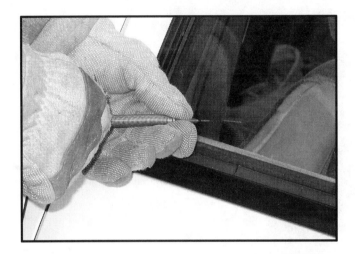

This center punch is a low-tech, simple tool with few moving parts. The tool consists of three parts: a punch, a spring, and a striker. To break tempered glass, the punch is positioned against the glass in a corner.

The striker is drawn back, which in turn puts tension on the spring.

The striker is released, strikes the punch, and breaks the glass.

Procedure 2
Removing Tempered Glass with a Halligan

Without a window punch, a Halligan may need to be used to take out tempered glass. If a Halligan must be used to remove glass, it should be positioned in a corner of the window so the handle will strike metal instead of entering the occupant compartment if swung with too much force. For best control, one crewmember can hold a Halligan while another crewmember strikes it with a sledge hammer or flat head ax.

If the center of the glass is broken with the Halligan, the crewmember on the tool will have a hard time stopping it before it enters the occupant area.

Without using the metal window frame or body of the vehicle as a stop, unnecessary injuries could occur.

Door and Side Procedures

Procedure 3
Creating a Purchase Point with a Halligan

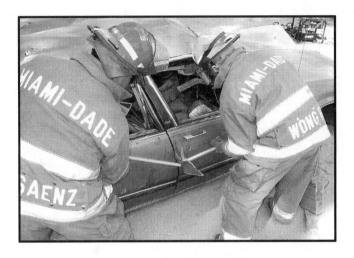

Creating a good purchase point for hydraulic spreaders is a simple procedure. One crewmember positions the adz of the Halligan over the gap between the front door and rear door about half way up the door. This position will provide good access to the door latch when the spreaders are put in position. The Halligan is struck, driving the adz into the gap.

The crewmember on the Halligan lifts up, then pushes down on the tool to spread the metal apart. Pulling out on the Halligan will do little to increase the size of the gap and is usually ineffective.

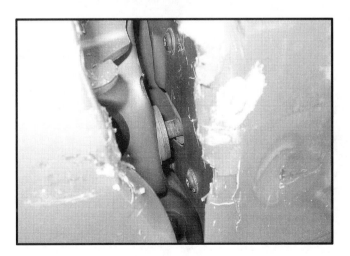

After working the tool up and down a couple of times, the Halligan is removed. The photograph shows what the spreader operator would see when positioning the tips to pop the door open.

Procedure 4
The *Pinch and Curl* Purchase Point

When the gap between a door and a pillar is too small to insert both spreader tips, the pinch and curl procedure can be performed. The first part of the procedure—pinching—is done by opening the spreader tips a couple of inches, then inserting one tip in the gap.

The tips are then closed firmly, pinching the sheet metal between the spreader tips.

With the tips closed firmly, the spreader is pivoted toward the front of the vehicle to bend the sheet metal. If the first attempt resulted in a gap that is still too small, the procedure is repeated until the gap is large enough to accommodate a good bite with the spreaders.

Door and Side Procedures

Procedure 5
Popping a Door Open with an Air Chisel

When it is necessary to pop a door open, high-performance air chisels provide a good backup to hydraulic spreaders. To cut through a door latch striker, the air chisel must be designed to operate in the 250–280 psi range and must have a sharp curved chisel bit. An examination of the gap between the door and the pillar will determine if the bolt is accessible to the air chisel.

If the gap is not quite big enough, a Halligan can be used to open up the gap. If the Halligan is unavailable or there are no other crewmembers available to help, the tool operator can use the air chisel to open up the gap.

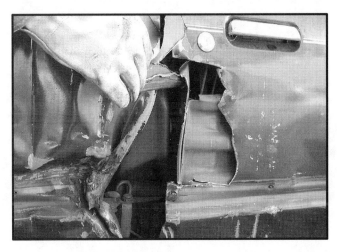

With a little work, often the striker can be exposed.

Procedure 5 *(continued)*
Popping a Door Open with an Air Chisel

To keep control of the air chisel, the operator should get in a position where forward pressure can be kept on the gun. If enough forward pressure is not maintained, the bit is likely to slip off the striker.

After 30 seconds of steady cutting, the chisel should be withdrawn and the bolt examined for progress. Occasionally a bolt will be so hard that it will resist the forces of the air chisel. If little progress has been made, the chisel bit should be examined for sharpness. A dull bit won't cut a striker and should be replaced. If the tool operator hears a sound similar to a rattling while cutting, it is an indication that not enough forward pressure is being exerted on the gun. If everything is working properly, the striker should shear off within a minute or so.

In some cases, the striker will be cut off; and in other cases, it will be forced out of the attachment hole by the powerful blows of the air chisel. In either case, the door comes open to provide access to the occupants.

Door and Side Procedures

Procedure 6
Popping a Door Open with Hydraulic Spreaders

Popping a door open is the most common extrication procedure performed by firefighters. To safely pop a door, the crewmember on the spreaders should be in a position that will limit the possibility of injury. Two of the simplest, most effective ways to prevent injuries are to use good body mechanics and to have the tool positioned between the operator and the door being popped open. If the operator stands between the tool and the door, there's a greater chance of the door popping open and striking the operator.

Getting a good purchase point is an important part of an effective operation. If the tips can't be inserted deep in the gap, the gap should be opened up. If a few seconds are spent at the beginning of the operation by creating a good purchase point, the metal will be less fatigued and less likely to tear. Tearing reduces the effectiveness of a spreading operation and should be avoided.

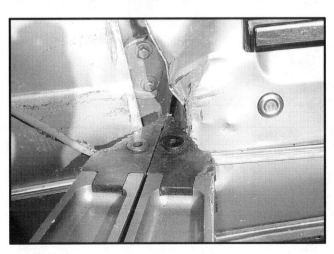

Once a good purchase point is opened up, the spreaders are inserted and opened. With experience, the tool operator will learn when the tip's grip on the door will hold and when it will tear loose. Before the tips lose their grip the tips should be closed and reinserted. The opening, closing, and repositioning of the tips will decrease the chance of tearing the metal.

Vehicle Extrication: A Practical Guide

Procedure 6 *(continued)*
Popping a Door Open with Hydraulic Spreaders

In the photograph, the metal in the rear door is starting to tear, reducing the effectiveness of the spread. One solution to this problem on a four-door sedan is to reposition the spreader tip against the rear door hinge. By design, the hinges and supporting hardware are stout and will often serve well as a push-off point for the spreaders.

Once the spreaders start to make progress, the spreaders may change position as they follow the failing door metal. Usually there is no problem when this occurs, but the tool operator should avoid becoming off balance in case the tips lose their grip.

As the latch on the door separates, the door will pop open. The tool operator is responsible for making sure the area is clear of all personnel before the door is popped open.

Door and Side Procedures

Procedure 7
Vertical Spread to Pop a Door Open

The deep V shown in this door complicates any effort to pop the door open with hydraulic spreaders. If the spreaders are inserted at the latch and the force applied to the door, the V would continue to deepen, pushing in on the occupant.

Existing injuries of the left chest, abdomen, pelvis, and legs could be aggravated if the door were to be pushed deeper into the passenger compartment.

Instead of spreading at the latch, the spreaders can be inserted between the top and bottom of the window opening in an attempt to push the door outward. If possible, the tips are positioned so the upper tip is further into the vehicle than the bottom tip. If the tips are positioned one above the other, the door would be forced down, not outward. Achieving this optimum angle will require the spreader be positioned away from the deepest part of the V.

Vehicle Extrication: A Practical Guide

Procedure 7 *(continued)*
Vertical Spread to Pop a Door Open

During the first vertical spread, the top of the window frame may be resisting the spreader. The spreaders can be backed out, and the cutters brought in to cut the window frame out of the way.

With the window frame out of the way, the spreaders are used to make another spread. If the spreaders get a good bite on the door and the spread is going smoothly, the spreaders should be opened to their maximum opening distance.

As a result of the first one or two vertical spreads, the deep V will be pulled away from the inside of the passenger compartment. It may be possible to place the lower spreader tip directly on the bottom of the V.

Door and Side Procedures

Procedure 7 *(continued)*
Vertical Spread to Pop a Door Open

In some cases, the door will unlatch during the vertical spreading steps. If the spreaders have met their maximum opening distance and the door has not popped open, the spreaders can be used for a nearly vertical attack of the door latch.

The tool operator will have a good view of the latch and should be able to position the tips in an effective spot. When operating the spreaders in this position, the crewmember on the spreaders may be drawn closer to the door to reach the latch but should avoid standing in front of the door to avoid injury should the door pop open with force.

The metal part of the door has been forced open, leaving only the interior trim pieces in the opening. These can be forced out of the way by hand or spread out of the way with the spreader.

Procedure 8
Popping a Rear Door with Hydraulic Spreaders

Popping a rear door on a four-door automobile is similar to popping a front door except accessing the latch is a little more difficult.

A Halligan can be used to create the initial purchase point, but some firefighters prefer to use a *vertical crush*. To open up the gap at the latch, the spreaders are used to compress the door causing deformity in the form of a larger gap. To perform the vertical crush, the spreaders are positioned with one tip on the inside of the door and the other on the outside.

As the spreaders close, the gap at the latch increases in size. Repeated attempts at this procedure usually do not increase the gap, making it unnecessary to crush the door more than twice. If the gap is not large enough for the spreader tips, the *pinch and curl* method can be used to open it up.

Procedure 8 *(continued)*
Popping a Rear Door with Hydraulic Spreaders

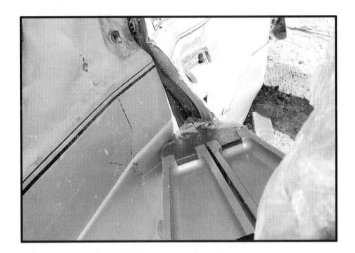

When positioning the tips of the spreaders to pop the rear door, it is difficult to see the latch assembly. This problem occurs because the latch is positioned several inches ahead of the rear of the door.

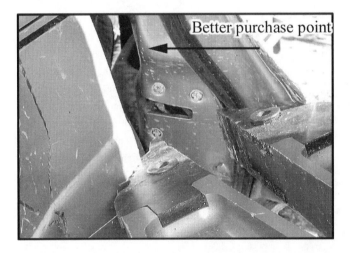

Multiple attempts to spread the door are often needed.

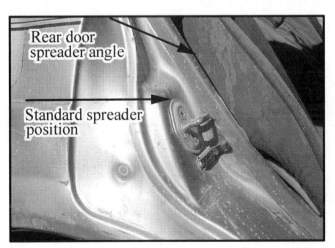

Examination of the door opening reveals the difference between the typical front door and rear door. The primary cause of difficulty is the inability to insert the spreader near the latch. One good solution is to change the positioning of the spreaders. After the door has spread open as far as possible, the spreaders can be repositioned above the latch and angled downward as shown in the photograph. This is often successful because more of the tip is inserted between the door and the body of the vehicle, creating a better bite and less slippage.

Vehicle Extrication: A Practical Guide

Procedure 9
Manually Widening a Door Opening

Doors that have been popped open with tools may not open wide enough to remove the victims from the vehicle. When tools aren't available to remove or spread the door open further, a few crewmembers may be able to smoothly push the door open wide enough to access and remove the victim. If not performed carefully, forcing the door open manually can result in the vehicle rocking, a situation that should be avoided. This is especially important for victims who may have received spinal injuries.

For maximum leverage, the crewmembers are positioned at the outer edge of the door. The door will act as a large lever that will fail (the desired effect) when overloaded. To avoid rocking the vehicle and aggravating a victim's injuries, one of the crewmembers coordinates the action. Smoothness and power are the key components of a successful widening of the door. Over-exertion should be avoided to prevent injuries to the crewmembers.

The door metal or hinge metal will tear or deform as the door is pushed toward the front fender.

Door and Side Procedures

Procedure 10
Widening a Door Opening with a Come-along

After a door is opened, there may not be enough unobstructed space to access and remove the victim. If attempts to manually force the door open fail, a come-along can be used to open the area. This come-along procedure can be done by one well-trained person, but, to be done quickly, three crewmembers are needed. To get started, the officer directs one crewmember to rig the door while another crewmember rigs an anchor chain on the chassis.

The door rigger brings a come-along and a chain to the area adjacent to the door. The door rigger places the sling hook end of the chain in the hand closest to the hinge. The other hand grasps the chain about 5 ft from the hook. The hook is then passed through the window opening and is positioned at a point in the middle of the door.

The hand closest to the latch then swings the chain under the door and towards the hook.

163

Procedure 10 *(continued)*
Widening a Door Opening with a Come-along

While the sling hook is held firmly, the chain is snapped through the latch. If the sling hook is not equipped with a latch, the rigger should position the hook with the open part of the hook facing away from the door. This will result in the chain being placed under a load in the back or the thickest part of the hook.

Once the chain is in position in the sling hook, the tool operator maintains light tension on the chain. Keeping light tension prevents the chain from falling out of the hook, holds the rigging in position, and makes it easier to use the chain shortener. While keeping tension on the chain, the rigger reaches down to the ground, picks up the chain shortener, and attaches it to the chain in the area of the end of the door.

Door and Side Procedures

Procedure 10 *(continued)*
Widening a Door Opening with a Come-along

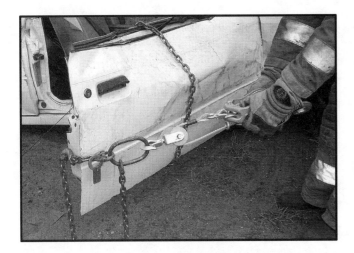

Once the chain shortener is attached, the rigger continues to hold tension on the chain to prevent it from coming apart. While holding the master link in one hand, the rigger reaches down and picks up the block of the come-along with the other hand. The hook on the block is attached to the master link, all the while keeping tension on the system. The come-along is then walked to the anchor chain.

The anchor chain is attached to a strong point on the chassis. The chain is shortened so the master link is positioned just above the top edge of the hood. After hooking the come-along to the master link of the anchor chain, cribbing is placed between the chain and the hood and grill area. The officer then directs the come-along operator to take the slack out of the chain and for the crew to check the rigging.

If the rigging is correct, the crewmember operates the come-along until the required space is created. The operation is stopped before the come-along slips off the front corner of the hood.

Procedure 11
Removing a Door with an Impact Wrench

Hinges can be attached to a vehicle in a variety of patterns and configurations. On occasion, the hinge will be attached so that all the bolts are exposed if the fender is moved out of the way during or after the crash.

If all the bolts are exposed, an impact wrench can be used to remove the bolts quickly.

The use of an electric, battery-powered, or pneumatic impact wrench not only provides a quick method of removal, but it frees up the hydraulic tools so they can be used in another area.

Door and Side Procedures

Procedure 12
Removing a Door at the Hinges with an Air Chisel

An air chisel equipped with a sharp, curved bit can be used to remove a door by cutting through the hinges. When deciding how to cut a hinge, most would think that the best approach would be to cut the pin that connects the halves of the hinge. Unfortunately, attempts to cut the pin usually result in the chisel getting stuck between the two halves of the hinge. A better approach is to cut through the hinge itself.

The tool operator will need to be in a good position to apply some force behind the gun. To make the cut, an air chisel designed to operate in the 250–280 psi range will be needed along with a sharp curved chisel bit.

Some hinges are tougher than others, and every so often, a hinge will be encountered that will be difficult or nearly impossible to cut. These hinges however, are in the minority not the majority. The progress of the cut in the photograph is after about 20 seconds of cutting.

Vehicle Extrication: A Practical Guide

Procedure 12 *(continued)*
Removing a Door at the Hinges with an Air Chisel

After the top of the hinge is cut, the bottom half can be started. When selecting the path of the cut, the shortest, least complicated route should be chosen.

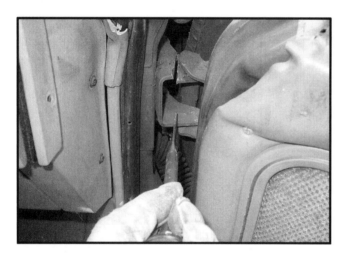

The second cut is lined up on the left half of the hinge and is planned to proceed through the shortest distance possible. The right half of the hinge will be missed if the shape of the bottom half is similar to the top half.

The line up was correct. The right half of the hinge was only nicked and had no impact on the speed of the cut.

Door and Side Procedures

Procedure 12 *(continued)*
Removing a Door at the Hinges with an Air Chisel

With the top hinge cut, the space between the door and the hinge column can be opened up a little by applying downward pressure on the outer edge of the door.

This procedure will give the tool operator a little more room to work.

Deciding on the best spot to cut is a little more complicated on the lower hinge. A quick examination will reveal how the parts are connected and how they can be cut for maximum effectiveness.

Procedure 12 *(continued)*
Removing a Door at the Hinges with an Air Chisel

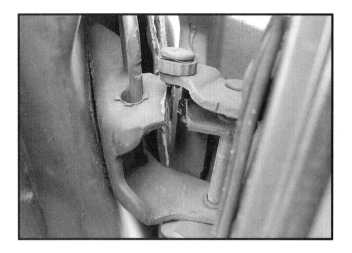

With the top half of the bottom hinge cut, the bottom half can be cut and the door removed.

As the cut is completed, the tool operator should take steps to stay clear of the door as it falls or have another crewmember hold it as the hinge is severed. These four cuts required less than one air cylinder to complete.

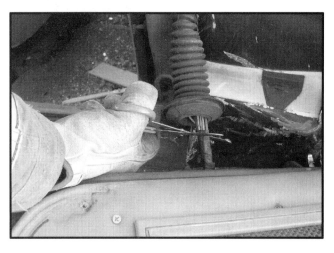

To move the door out of the work area, the wiring harness is cut.

Door and Side Procedures

Procedure 13
Removing a Door with Spreaders—Exterior Approach

After a door has been popped open, the spreaders can be used to remove the door at the hinges. While not as effective as using cutters for the task, spreaders may be the best choice when the cutters do not have the power needed to make the cut or when the area of the hinge is too tight for the cutters. When using the spreaders, the goal is to apply spreading force between the door and the front hinge pillar. Usually the best approach is to start with the bottom hinge by placing the spreaders tips on top of the hinge. A 4X4 or backboard is placed between the door and the "B" post to prevent the door from closing and having to be popped open again with the spreaders.

To gain access to the hinge, it may be necessary to open up the space so the tips can be inserted. This is done by inserting the tips as far as possible then opening. The sheet metal of the fender and door will pose no challenge to the spreaders.

As the spreaders are opened, the tool operator may see that the spreaders have made a solid contact and the hinges may start pulling apart. In this case, the tool operator should continue with the spread until the hinge comes apart. If the sheet metal starts to tear instead, the tips are repositioned directly on top of the hinge and the spread is started again.

Procedure 13 *(continued)*
Removing a Door with Spreaders—Exterior Approach

With the bottom hinge broken, the top hinge can be addressed. To create a purchase point when the sheet metal is overlapping and provides no space to insert the tips, the *pinch and curl* method can be used to create an opening for the tips. The tips are opened a couple of inches, then a tip is positioned on each side of the sheet metal. The tips are then closed on the metal, pinning it between the tips.

The tool operator then walks the spreaders toward the rear of the vehicle. The leverage created by the spreader causes the metal to bend easily.

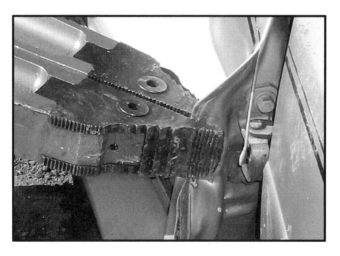

With the sheet metal out of the way, the hinges are exposed.

Door and Side Procedures

Procedure 13 *(continued)*
Removing a Door with Spreaders—Exterior Approach

With good access created, the tips are placed on top of the upper hinge.

As with the bottom hinge, the initial spreading of the door may be solid enough that the sheet metal stays intact and the hinge starts to come apart.

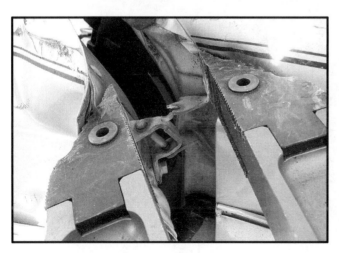

The hinge pin in this hinge tore through the upper portion of the hinge then through the bottom, separating the door from the hinge pillar.

Vehicle Extrication: A Practical Guide

Procedure 13 *(continued)*
Removing a Door with Spreaders—Exterior Approach

During the spreading operation, the tool operator will be concentrating on the hinge but should also be aware of the movement of the door. As the upper hinge is pulled apart, the door will fall free, held only by the wiring harness.

This lightweight door fell next to the tool operator's leg without making any significant contact, but a larger, heavier door could cause some problems if the tool operator were in the way.

The wiring harness between the hinge column and the door can be cut with a set of cutters or bolt cutters after pulling outward on the door to make the wires taut.

Door and Side Procedures

Procedure 14
Removing a Door with Spreaders from the Interior

Removing a door with spreaders from the interior is an option that can be utilized when the hinges are not accessible from the exterior of the vehicle. For the spreaders to be effective, the hinge pillar must have considerable strength to resist the spreading force.

To position the tool properly, the tool operator will often have to lean inside the vehicle. This can be a problem if the victim is seated next to the door that is to be removed.

If the decision is made to approach the hinges from the interior, the tips of the spreader should be positioned on top of the bottom hinge if possible.

Vehicle Extrication: A Practical Guide

Procedure 14 *(continued)*
Removing a Door with Spreaders from the Interior

It is the tool operator's responsibility to ensure that there are no crewmembers in the path of the door when either hinge is broken. While most doors are not propelled very far when the hinge breaks free, occasionally a door will fly a couple of feet away from the vehicle. After the area is cleared of crewmembers, the spreaders are opened until the door breaks free.

With the bottom hinge broken in half or torn out of the hinge pillar, the top hinge is approached. Like the bottom hinge, it is best to lay the spreader tips on top of the hinge when attempting to break the hinge.

When both hinges are broken, the door will spring free. As the door is being broken free, it may start to twist or turn in a direction that could cause it to strike the tool operator. The tool operator should be aware of the location and movement of the door and take steps to avoid being struck by it when it falls. This may mean simply repositioning or getting another crewmember to help hold the door. After the door breaks free, the wires between the door and the hinge pillar are cut with heavy-duty EMS shears.

Door and Side Procedures

Procedure 15
Removing a Door from the Interior with Cutters

After a door is popped and swung open on its hinges, crewmembers may need even more space to work than is provided by the open door. If the door is to be removed at the hinges, a quick examination of the hinge area will be needed to determine if it is best to approach the hinges from the inside or outside of the door. The hinges on the front door of this vehicle are hidden from the outside, making an outside approach a poor choice.

With the door swung all the way open, the hinges are in plain view and accessible from the inside. A quick look at the hinges is all that is needed to find the section of the hinge that contains the least amount of metal that will need to be cut. When cutting hinges, usually there is not much choice in positioning of the tool because of the limited amount of unobstructed space around the hinge, but it is still worth a look in case there is room for positioning the tool.

After a size-up of the hinges, the cutters are positioned on the hinge. If there is enough room to precisely position the cutters, the hinges should be positioned as deep as possible into the cutter blades to maximize the cutting power of the tool. If there is room for adjustment, the cutters should be positioned to cut through the smallest amount of metal possible.

Vehicle Extrication: A Practical Guide

Procedure 15 *(continued)*
Removing a Door from the Interior with Cutters

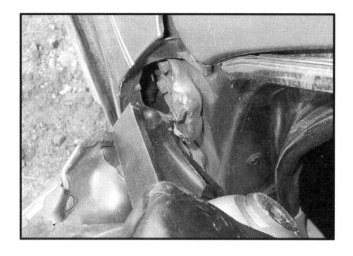

As the cutter shears through the hinge, it will encounter some weak parts and some strong parts of the hinge. This strength differential may cause the cutters to twist and turn as it cuts. This twisting action is common and is acceptable as long as the operator is not placed in danger by the movement of the tool and as long as the cutter blades do not start to separate as they cut. When cutter blades close, only the smallest gap should exist between the blades. If this gap is enlarged, the object being cut can get wedged between the blades, forcing them apart even further. If allowed to continue, this action can lead to a fracture of the blades.

As the hinges are cut, the door may start to move in response to the forces being applied to the hinge. The tool operator will be focused on the hinge and would benefit from a little assistance from other crewmembers to control the door.

After the second hinge is cut, the door can be pulled away from the vehicle to put tension on the wiring harness that is connected to the door. Once taut, the cutter can be used to cut the wires. If the wires are not pulled taut, instead of being cut they may simply get wedged between the cutter blades as they close.

Door and Side Procedures

Procedure 16
Removing a Door from the Exterior with Cutters

If examination of the door reveals that the hinges are accessible from the outside, they can be cut from that position. The top hinge is visible from the outside, but the bottom hinge is hidden from view. In this situation, the tool operator should cut what can be seen and then try to manipulate the door to expose the second hinge. After the first hinge is cut, the best approach for the bottom hinge can be determined.

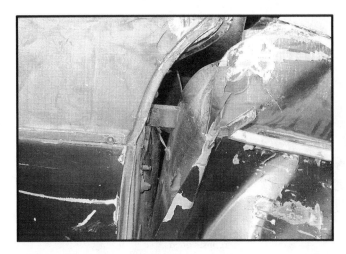

This hinge is simple in design but stout. The primary factor in cutting through tough hinges is to get them as deep as possible in the cutter blades.

With a little manipulation and fine adjustments, the cutters are positioned on the hinge.

179

Vehicle Extrication: A Practical Guide

Procedure 16 *(continued)*
Removing a Door from the Exterior with Cutters

When viewed from the inside, it is clear that the tool operator has a choice of two approaches to the hinge; coming in from above or coming in from the side.

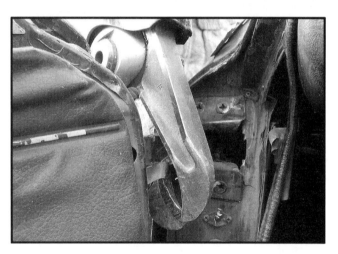

The approach from above may seem like the best choice because of the easy access and the ability to position the hinge deep in the cutter opening. The drawback to this approach is the increased amount of contact between the cutter edges and the hinge. The large amount of contact can actually reduce the ability to cut the hinge because the force is being distributed over a larger area. The final determining factors will be the cutting force of the cutters, the mass of the hinge, and material used to make the hinge. If the cutters won't cut the hinge in this position, they can be repositioned.

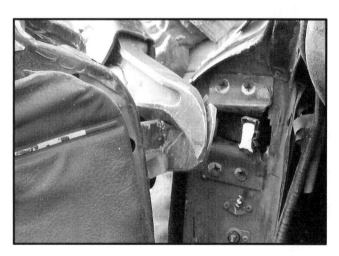

By repositioning the cutters, less of the hinge will make contact with the blades resulting in concentration of the force. While the hinge is not positioned as deep in the cutter, this position often proves to be a more effective approach.

Door and Side Procedures

Procedure 17
Popping and Removing a Van Door

Rear doors on vans in the United States are usually on the curb/right side of the vehicle. Opening these doors may provide the only means of access to multiple rear seats and passengers. Wide variations in the design of these doors require a quick examination to determine the best way to open them up.

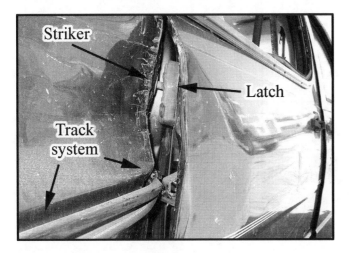

The gap at the rear of this sliding door reveals a latch device and track system that will need to be spread or cut out of the way. Multiple latches may be encountered when opening these types of doors.

With good access, hydraulic cutters can be used to cut the striker and the track roller.

Vehicle Extrication: A Practical Guide

Procedure 17 *(continued)*
Popping and Removing a Van Door

After the striker is cut, the spreaders are used to separate the door from the body of the vehicle. In some cases, it may be necessary to cut, spread, and cut again to release the door.

With the rear of the door released from the body of the van, the front half is evaluated to determine how best to release it. The van in the photo utilized a track and roller system in the front of the door that was accessible and easily cut with the hydraulic cutters.

To reduce the chance of the door falling and causing injuries, the door should be supported and lowered to the ground by another crewmember.

Door and Side Procedures

Procedure 18
Creating a Third Door with an Air Chisel

Gaining access and removal of rear seat passengers of a two door vehicle is always tough. One solution is to remove the side of the vehicle in a procedure that is generally referred to as a *third door*. In essence, the latch pillar and the metal around it is removed creating a new opening.

The key to successful completion of this procedure is the creation of a large inspection hole in the sheet metal. The inspection hole exposes the interior of the structure for both size-up and access for cutting. To create the inspection hole, a cut is made an inch or so under the window, extending from the rearmost portion of the window to an inch from the edge of the latch pillar. The inch spacing is intended to avoid reinforced, double layers of metal that will slow the process.

The cut proceeds downward to a point even with the front door rocker panel.

Procedure 18 *(continued)*
Creating a Third Door with an Air Chisel

As the air chisel bit approaches the level of the rocker panel, the tool operator can make a turn with a chisel and continue horizontally. The other option is to stop, reposition for the horizontal cut, and start again. If possible, the operator should opt for making the turn instead of stopping and repositioning. If the operator stops and repositions, it will may be a little difficult to start the new cut because the metal will be weakened and more likely to bounce around instead of cutting.

The next cut starts at the original beginning point and proceeds downward just behind the plane of the seat back. When making this cut, it is easier to work downward than trying to work the bit upward from the end of the previous cut. As the wheel opening is approached, the operator starts turning and cuts along the shape of the wheel opening. Again, an inch or so away from the edge is adequate to avoid reinforcement and sheet metal overlaps.

The cut continues until the other cut is met.

Door and Side Procedures

Procedure 18 *(continued)*
Creating a Third Door with an Air Chisel

With the cuts completed, the outer layer of sheet metal is removed to expose the inner supports and structure. With a good view and access, the tool operator can create a plan of how the remaining structure can be removed. When sizing up the metal, the operator should look for any holes made during construction of the vehicle and utilize them in a manner that eliminates some cuts. This is a process of *connecting the dots*.

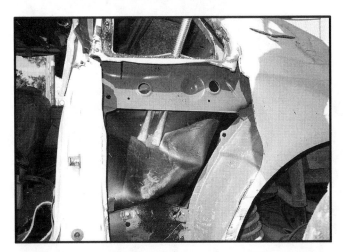

To open up the structure some more, the inner panels are removed. Usually, they can be pulled out manually with a sharp pull but may require some prying with a Halligan. These decorative panels are often held in place with plastic buttons or wedge-like plastic attachments referred to as *Christmas trees*.

After initially being pried away with a Halligan, the panel is manually removed. Brute force is usually the best way to remove the panel because it may come out cleanly without any jagged edges that may interfere later with victim removal.

Procedure 18 *(continued)*
Creating a Third Door with an Air Chisel

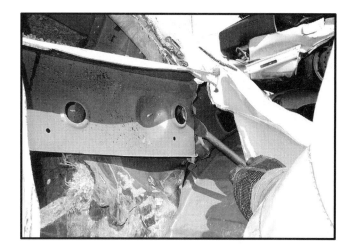

There are three areas remaining that need to be cut: the top of the "B" post (if present), the bottom of the latch pillar, and the horizontal member under the window. The horizontal window member is cut in an area rearward of the seat back. A planned series of cuts are made to sever this component. If a T or dual chisel bit has been used to this point, it should be switched at this time to a curved chisel bit.

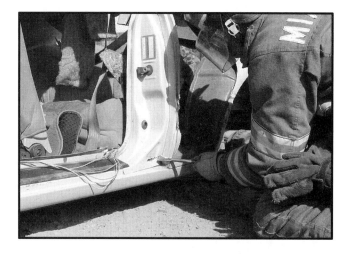

The perimeter of the latch pillar is cut, extending from the inspection hole around the pillar to the inside.

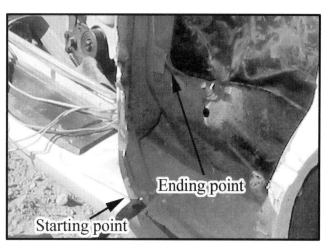

The advantage of making the inspection hole can be seen here. Most of the latch pillar can be cut from the outside, but the inner portion can be cut more easily by accessing it through the inspection hole. Some vehicles have more metal in this structure than others. Instead of plastic interior trim, sheet metal may be found inside the inspection hole. In this case, the cut at the rear of the window is extended all the way down to the latch pillar cut. The cut should conform to the shape of the seat to avoid jagged, protruding edges.

Door and Side Procedures

Procedure 18 *(continued)*
Creating a Third Door with an Air Chisel

The "B" post will be cut high, so the seatbelt should be cut to allow easy movement of the post after it is cut. A hazard size-up should be done before the "B" post is cut. The process for the hazard assessment is explained in chapter 5.

With the hazard assessment completed, the "B" post is cut as high as possible, avoiding the shoulder harness assembly and its thick backing plate. To avoid having the curved chisel bit getting stuck in the post, only the inner half of the blade should be used to cut around the perimeter of the post. Once the entire perimeter is cut, the curved cutter bit can be used to cut through any remaining metal in the center of the post. If the chisel bit gets stuck and can not be pulled free, the bit is released from the gun and replaced with another bit. The new bit is used cut out the stuck bit and to finish the "B" post cut.

If all the cuts were made in a systematic order, the entire severed structural member can be removed. Any remaining hidden pieces of attached metal that may prevent the removal of the member can be identified by twisting and pulling. The air chisel is then used to sever them. Jagged edges will be present but should be recessed behind the level of the seat bottom and back if the cuts were made correctly. Perfect cuts often are not possible, so the jagged edges should be covered, if possible, to protect the victim during removal.

Procedure 19
Creating a Third Door with Hydraulic Tools

When using hydraulic tools to perform a third door, relief cuts are made to weaken the metal so it can be bent out of the way. In addition to the two relief cuts required for this procedure, the "B" post must be severed or removed. If the roof has been removed, the cut will probably have been made low on the post to remove the post as an obstacle to accessing the victim. If the roof has not been removed, the "B" post can be cut high and used for leverage when the side is pulled down.

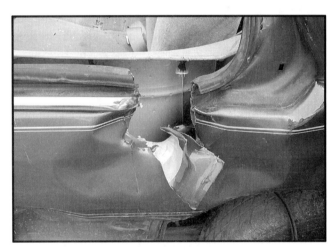

The relief cut can consist of a single cut, or can be made larger by making two cuts then prying the sheet metal down. If a single cut is made, there is a chance the metal may bind when it is being bent downward. By making two cuts, there is less chance of a bind. In this case, cutter blades were not large enough to include the interior plastic in the cut. This plastic can either be pulled up and out of its attachment points or bent down with the metal.

To weaken the vertical structure a horizontal relief cut is made into the bottom of the latch pillar.

Door and Side Procedures

Procedure 19 *(continued)*
Creating a Third Door with Hydraulic Tools

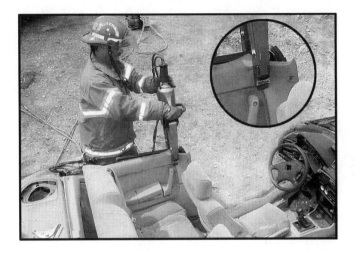

With the three structural areas that provide strength to the body severed, the spreaders are clamped down on the upper forward corner of the structure. For maximum leverage, only tips should be making contact with the metal. The farther forward the tips are positioned, the greater the leverage when pulling down.

After the spreaders have been clamped down on the metal, the tool operator pulls outward and downward, forming a crease that runs from one relief cut to the other. The spreaders may need to be repositioned for maximum leverage, based on how the crease develops.

As the metal bends at the crease, some resistance may be encountered, especially in larger, sturdier vehicles. To further weaken the structures providing the resistance, the original relief cuts can be expanded. By combining additional cuts and bending, good access to the rear seat occupant can be achieved.

Procedure 19 *(continued)*
Creating a Third Door with Hydraulic Tools

If the size and strength of the vehicle prevents the third door from being pulled down, rams can be used with good results. Selection of the ram should be based on the available space to position the ram. The longest ram that fits in the available space should be used. If the "A" post is intact, it can be used as a push-off point for the base of the ram.

The ram positioned so the third door area is pushed down and out.

With an angle set that will push the metal outward, there is less chance of the metal bending in some unanticipated manner that results in movement into the passenger compartment.

Door and Side Procedures

Procedure 19 *(continued)*
Creating a Third Door with Hydraulic Tools

A firm grip on the ram will be needed in the event the head of the ram slips off the metal and trim as the push is completed.

By positioning the base of the ram high on the "A" post, a crease is created between the two relief cuts.

When the ram is fully extended or when position of the ram must be changed to continue the push, a lower point can be used. The initial extension of the ram creates the crease and defeats most of the resistance making it easier to use the ram in a straight front to rear angle.

Procedure 20
Creating a Fourth Door with an Air Chisel

Many vans in the United States are equipped with a rear door on the curb or passenger side only. Should access to the street or left side of the van be necessary, the side can be removed with an air chisel. To remove the side of the van, the sheet metal skin and underlying support structure will have to be removed. While the procedure may seem complicated, a skilled air chisel operator can complete the operation in less than five minutes. The procedure consists of three steps. The first involves removing a large section of the outer sheet metal to provide access to the inner layers. The first cut is made parallel to and below the window opening.

The next cut is a continuation of the first cut and extends from the bottom of the window to floor level.

The rearmost cut is made behind the seat back. If the cut were in front of the seat back, it would unnecessarily impede access to the seat area. The gas filler hose is avoided when making the cut.

Door and Side Procedures

Procedure 20 *(continued)*
Creating a Fourth Door with an Air Chisel

The fourth cut is made along the bottom at the level of the floor. If made any lower, additional structures may be encountered that would delay completion of the cut and not add any real benefit. When making the fourth cut, the flap of sheet metal can be folded down providing access to the inner surface of the sheet metal. The condition and shape of the damaged metal will determine the best approach to be used.

With the outer layer of sheet metal removed, the underlying structure can be observed, and the order of the next few cuts can be planned. To provide a better view of the structure, the interior panel can be removed.

The interior panels can be made of a variety of materials, all of them fairly lightweight. Some can be removed by simply tearing them out by hand while others will need to be cut out. Forcefully removing by hand should be attempted first as the entire piece will be removed, eliminating a lot of cutting time and jagged edges.

Vehicle Extrication: A Practical Guide

Procedure 20 *(continued)*
Creating a Fourth Door with an Air Chisel

With both the exterior sheet metal and the interior panel removed, the remaining structure can be evaluated and the cuts planned.

For maximum benefit, the cuts should be made so the greatest amount of material can be removed with the fewest cuts. The locations of the cuts are determined by the access created when the outer layer was removed. It is important not to get bogged down by trying to cut in areas where the outer layer has not been removed. If it is necessary to cut in unexposed areas, it is best to make additional cuts to the outer layer to open it up before proceeding to the inner layers.

The curved chisel bit is the best choice for cutting through these structural members. Care must be taken not to slip off the metal with the bit and injure the victim. A systematic approach is best when making these types of cuts, starting at one end and working as far as possible before repositioning. Jumping around with the air chisel will often result in little tabs of metal holding the member together when it appears the cut is completed.

Door and Side Procedures

Procedure 20 *(continued)*
Creating a Fourth Door with an Air Chisel

A decision must be made whether to remove the entire "B" post and striker section or leave it intact and simply remove the horizontal member beneath the window. This decision will be based on the amount of access needed, the post-crash configuration of the structure, and the time needed to complete the operation. In this example, the horizontal member is to be removed, leaving the "B" post intact.

With the horizontal member removed, any additional pieces of metal in the way can be cut away or bent out the way.

The completed fourth door evolution provides good access to the rear seat occupant. If the arm rest does not fold up and out of the way, it can be cut with the hydraulic cutters.

Vehicle Extrication: A Practical Guide

Procedure 21
Side-Out

The side-out procedure is a good choice when a four-door vehicle has sustained moderate to severe damage to the center pillar and "B" post. Vehicles with this type of damage are poor candidates for the conventional approach of popping the front door to remove victims. While the door can be popped in the routine manner, often there are difficulties created as the door moves into the passenger compartment.

The side-out procedure involves the removal of the front door, rear door, "B" post, and center pillar, all in one piece. This procedure will involve cutting the "B" post, so the vehicle will need to be evaluated for hazards associated with these cuts. The procedure for conducting a hazard analysis is detailed in chapter 5. To get started with the side-out, the rear door is popped open, while leaving the front door latched.

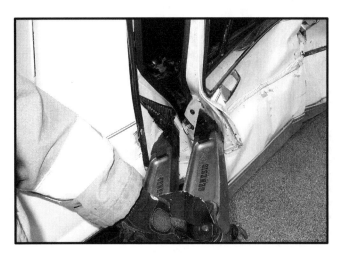

Popping a rear door of a four-door vehicle is a little more difficult than popping a front door because of the shape of the rear latch pillar and striker. For best results, positioning the spreader a little higher than normal is recommended.

Door and Side Procedures

Procedure 21 *(continued)*
Side-Out

After doing a quick hazard check as described in chapter 5, the "B" post is cut high with hydraulic cutters. Just before positioning the cutters on the "B" post, the tool operator should reach inside to locate the shoulder harness assembly. The assemblies have backing plates that will slow down all cutters and cause problems that can be avoided.

These cutters are just above the shoulder harness backing plate that is inside the "B" post.

At this point, the rear door is open and the "B" post is cut, disconnecting it from the roof. The next step is the severing of the center pillar from the rocker panel. Ideally, the center pillar could be cut with one cut, but that is seldom, if ever, the case. The rear portion of the center pillar is exposed and accessible because the rear door is open, but the front portion of the center pillar remains hidden by the front door. To solve this problem, the cutters are first used to make a cut as deep as possible in the center pillar.

Vehicle Extrication: A Practical Guide

Procedure 21 (continued)
Side-Out

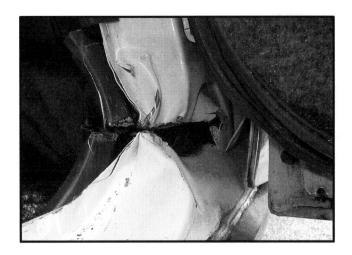

The cut is made as deep as possible beneath the bottom hinge. The cut will be the start of a metal tear that will continue until the center pillar is completely separated from the rocker panel. It is important that the cut be horizontal and not down into the rocker panel.

To start the tearing motion, a spreader is used between the forward lower corner of the rear door and the rocker panel. Positioning the spreaders correctly can be a little tricky but can be made easier by an understanding of how the tear progresses.

The metal tears because the center pillar is placed in tension. Tension means an object is being pulled in opposite directions. To achieve tension, the upper part of the center pillar must be pulled upward while the lower part of the center pillar is pushed downward. The cut that was made by the cutters would make it easy for the tear to begin. However, strictly vertical movement of the pillar would be difficult, because the "B" post would strike the roof if pushed upward.

Door and Side Procedures

Procedure 21 *(continued)*
Side-Out

To avoid bumping the "B" post into the roof, the "B" post and the center pillar assembly are pushed outward and upward at the same time. To make that happen, the spreaders are positioned at about 45 degrees between the rocker panel and the front, lower corner of the rear door. This will push the "B" post upward and away from the vehicle at the same time. This photograph was shot at the level of the rocker panel, looking upward.

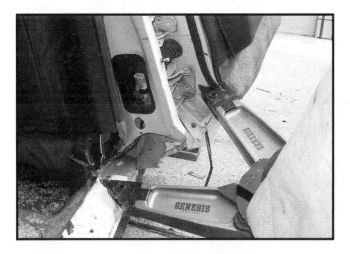

As the spreader is opened, the rear door will pull the hinges, which in turn pull on the center pillar. The relief cut made by the cutters is the starting point of a tear that continues until the entire center pillar is severed. As the metal tears, the spreader may require repositioning, but that is not usually the case.

As the spread is made and the tear continues, the rear door will move upward and outward until the tear is completed. The tool operator has the responsibility of clearing personnel from the area adjacent to the doors before the tear is completed and the door swings open.

Vehicle Extrication: A Practical Guide

Procedure 21 *(continued)*
Side-Out

As the tear is completed, any wires connecting the center pillar and the rocker panel are cut with EMS shears or side cutters.

The end result of the spreading operation can be seen. As the rear door was spread away, the front of the center pillar tore cleanly.

After the center pillar wires are cut, the two doors and center pillar are swung open like one giant door. Once open, the front door can be removed with the spreader. If possible the bottom hinge is broken first, followed by the top hinge.

Door and Side Procedures

Procedure 21 *(continued)*
Side-Out

Shears can be used instead of spreaders if access to the hinge is possible.

The combined weight of the two doors and center pillar can be substantial. To reduce the chance of injury to the tool operator, a crewmember helps support and control the door as it is cut free. The crewmember helping should be careful of the pinch point between the doors and keep hands clear of the area if possible.

The side-out procedure creates unobstructed access to both front and rear seat passengers. This procedure can be very valuable when the crash has caused the upper half of the front seat to be pushed into the rear half of the vehicle. That type of situation often results in the victim being in both the front and rear halves of the vehicle with the center pillar preventing easy removal of the victim.

Procedure 21 *(continued)*
Side-Out

There is a situation that can complicate the side-out procedure. If the front of the center pillar fails to tear in the manner desired, the tear may bypass the front of the pillar and extend into front door rocker panel.

Fortunately the complication is minor as the sheet metal of the rocker panel will be accessible.

Instead of allowing the tear to continue to the front pillar, shears are used to cut the rocker panel metal, releasing the doors and center pillar.

Door and Side Procedures

Procedure 22
Creating a Fourth Door with Hydraulic Tools

Access to rear seat passengers on the driver's side of a van can be accomplished with hydraulic tools in a manner similar to the procedure used for sedans. The basic process is the same, only the shape and size of the metal is different.

The driver's door is first opened to provide access to the latch pillar. The latch pillar is then quickly evaluated for hazards as described in chapter 5. A relief cut is made in the latch pillar to start the weakening process. The cut is made parallel to the ground and as low as possible. If the shoulder/seat belt assembly is in the area of the cut, it should be avoided by going a little above or below the assembly.

The "B" post is severed completely after the hazard check. For large posts, the cut can be made from multiple directions to completely sever it.

Procedure 22 *(continued)*
Creating a Fourth Door with Hydraulic Tools

A vertical relief cut is made to weaken the structure beneath the window. The location of the relief cut should be made after visualizing the creation of the crease between the low relief cut and the high relief cut. Ideally, the high relief cut would be made behind the seat back of the rear seat, but that may not be effective if the distance between the cuts is so long that the crease won't form properly.

A ram is positioned to push the side of the van out and away. The "A" post can be used as a push off point if it's intact and sturdy. If the "A" post won't work, the door hinges can provide a sturdy push off point, but the angle between the latch pillar and the hinges may push the side of the van inward instead of outward. While not the best choice, the dashboard may provide a good angle and can be made sturdier with some cribbing.

This view from the "A" post area shows the optimum pushing angle for the ram. If a good angle can't be achieved, another option is to have a crewmember grab the top of the "B" post and pull it outward to help the metal get started moving in the right direction.

Door and Side Procedures

Procedure 22 *(continued)*
Creating a Fourth Door with Hydraulic Tools

If the dashboard is selected for the push-off point, the steering column area may provide enough support that cribbing may not be needed.

After the ram is fully extended, it can be retracted and repositioned to the "A" post, front hinge pillar, or hinges. By relocating the ram, a few extra inches of push can be obtained.

After the metal is bent down, any plastic trim that would hinder removal of the victim can be manually pulled out of the way.

Procedure 23
Opening a Trunk with Hand Tools

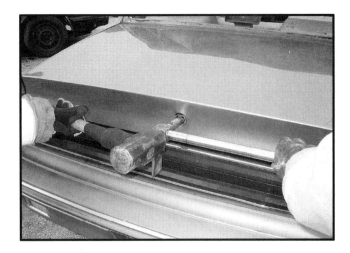

Crews may need to enter a trunk to access a battery or to release individuals locked in a trunk by accident or criminal intent. This procedure is used for standard, low tech trunk lids that open with a key. The pike of the Halligan is placed over the lock and is then struck with an ax or a sledge hammer.

The Halligan should be struck hard enough to push the lock out of the sheet metal, but not so hard that the latch device behind it is damaged.

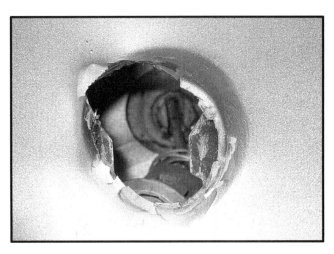

A flashlight is used to illuminate the lock, latch and metal connector that goes between the two parts. If the lock obstructs the view of the latch, it must be pried away to the side with a screwdriver. The metal connector may fall out of the latch, which simplifies the procedure.

Door and Side Procedures

Procedure 23 *(continued)*
Opening a Trunk with Hand Tools

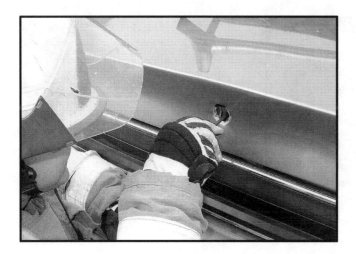

With the lock and the connector out of the way, a medium-size slot-head screwdriver is inserted into the hole. The end of the screwdriver should be oriented so the blade is vertical, not horizontal.

The target for the screwdriver appears as a vertical slot in the center of a circle. With the blade inserted in the slot, a quarter turn in the clockwise direction will unlatch the trunk lid.

The third photo on this page shows an upward view of the trunk latch with the screwdriver inserted in the slot.

Vehicle Extrication: A Practical Guide

Procedure 24
Opening a Trunk with Hydraulic Spreaders

Finding a good purchase point for the spreaders is made difficult by the overlapping trunk lid. To create a purchase point, the pinch and curl method can be used with good results. The spreaders are opened slightly, and then positioned so the trunk lid is between the open tips. The spreaders are then closed, getting a good bite on the trunk lid.

The spreaders are then lifted and rotated upward, causing the lip of the trunk lid to bend away from the body of the vehicle.

The spreaders are then repositioned so the tips are between the body of the vehicle and the trunk lid. The tips should be positioned to avoid damaging the fuel tank filler if it's positioned at the rear of the vehicle.

· 5 ·

Roof Procedures

Roof deformity, caused by a vehicle that has rolled over, come to rest on its side, or suffered a severe frontal or side collision, will often make extricating victims difficult for crew members. After crashes that cause minor roof damage, some roof procedures may be necessary to perform other non-roof procedures, like a dash roll-up. On other rare occasions, roof removal may be needed to ventilate the passenger compartment in order to lower the temperature or to eliminate gases from a ruptured battery. In many cases, roof procedures are necessary to simply see what is trapping the victim and to provide working room for tools.

Decisions have to be made about how much of the roof needs to be removed, which tools should be used, and which procedures will provide the best results. The selection of tools and procedures should result in the easiest, fastest, yet safest way to provide the desired work space. To make these decisions, it's helpful to break down the options and try to understand why one choice is better than another.

The Roof Structure

Roofs are built out of sheet metal; some of it is flat, and some of it is formed into tube-like shapes. The flat, or top part of the roof, is usually a single thickness of sheet metal, which is supported at all four edges with metal that's shaped to give it strength and rigidity. Roof posts are also shaped for strength and extend from the edges of the roof to sturdy parts of the body. To provide good rollover protection, roof posts may be reinforced. To give the flat part of the roof some strength, it's formed into a gentle arc and reinforced with one or two ribs that run from one side of the roof to the other. Occasionally, a roof has two layers of sheet metal separated by an inch or two of space.

Until recent years, roof procedures were simple operations that posed little danger other than exposing sharp edges after the metal was cut. That has changed a little with the installation of air bag systems that protect the occupant's head, neck, and torso. Like the more common air bags designed to protect drivers and front seat passengers during frontal crashes, the side-impact air bags and curtains inflate very quickly to protect the occupants. To inflate quickly, some side-impact air bag systems are inflated by one or more compressed gas cylinders that are hidden away in the edge of the roof or in the roof posts. The routine roof procedures have become more complicated because of concern about cutting through the compressed gas cylinders. The question then becomes, *how can a roof be cut if compressed gas cylinders are hidden in the roof structure, and how can an unintentional activation of the side-impact air curtain system be prevented?*

A Practical Approach for Sizing Up the Roof

To decide how to proceed with a roof operation, it is helpful to first clearly identify the needs and goals before trying to come up with the solutions. After the potential solutions are selected, potential problems can be identified. Following are the most important concerns regarding roof procedures:

- Activating unintentionally the side-impact protection system (SIPS)

- Puncturing (cutting) a compressed gas cylinder used in inflatable side-curtain systems

- Cutting the gas lifters used in hatchbacks

An unintentional activation of a SIPS during an extrication is unlikely, but history has shown that even well-designed products can perform in unanticipated ways. By acknowledging that vehicles are designed for occupants, not extrication crews, the potential for problems becomes evident. Fortunately, there are simple steps that can be taken by the crews on a crash scene to reduce the chance of failure and problems when working on a roof. These steps are becoming part of a new type of *universal precautions* to be used when working at crash scenes.

Roof Procedures

Universal Precautions for Crew Safety When Performing Roof Procedures

1. Assume that the vehicle has all types of undeployed air bags until proven differently. This includes air bags mounted in the steering wheel, dashboard, door, outer part of the seat back, and knee bolster as well as air curtains that drop down out of the ceiling at the side edges of the roof.

2. Turn off the ignition. This simple step does a lot of things to make the scene safer including removing power from most air bag systems.

3. Disconnect the battery. Disconnecting the battery helps eliminate the chance of a short circuit bypassing the ignition switch and other components that power the air bag system. In certain crashes, some BMW vehicles help by automatically severing the positive cable that's connected to the battery (see Fig. 5–5 and 5–6).

4. Try to stay clear of potential air bag inflation areas. To help estimate the air bag inflation area, the crew member should remember that air bags are designed to form a cushion between hard components in front of, and to the side of the occupants. By remembering that an air bag will fill the gap between the occupant and the metal, glass, or plastic, the area to be

Fig. 5–1 Deployed driver and front passenger air bags.

Fig. 5–2 Deployed driver, driver's torso, and front seat passenger air bags.

Fig. 5–3 A combination of multiple air bags including a drop-down inflatable curtain.

Vehicle Extrication: A Practical Guide

Fig. 5–4 Deployed door air bag.

Fig. 5–5 This BMW positive battery cable automatically disconnects when the air bags are deployed.

Fig. 5–6 The battery and automatic disconnect is located in the trunk.

avoided can be estimated. When cutting a roof or its posts, it may be possible to identify the presence of inflatable curtains. If inflatable curtains are present, there's a chance that they are inflated by a compressed gas cylinder.

5. Remove the interior trim pieces to expose any hidden, compressed gas cylinders. Extrication instructor (and author) Ron Moore has done a tremendous amount of research in the area of occupant safety systems and their impact on extrication operations. He was one of the first to recommend the removal of interior trim pieces to help locate the compressed gas cylinders used in the SIPS. Fortunately, the trim pieces that must be removed to expose these cylinders are usually plastic panels that are snapped into place and can be easily removed by hand.

6. Work around compressed gas cylinders, if present. This is another situation when clarity of thought is important. If the interior trim pieces are removed and compressed cylinders are located, that area shouldn't be cut. On the other hand, if the trim pieces have been removed and no compressed gas cylinders are found, a potential cutting location has been identified.

Roof Procedures

EXISTENCE AND NON-EXISTENCE OF AIR BAG LABELS

Careful observation of trim, seat, and glass may reveal labels indicating the presence of these systems. The labels often use abbreviations for the type of system, such as SRS (supplemental restraint system), SIPS (side-impact protection system), IC (inflatable curtain), among others. When unsure of an abbreviation on a trim piece, it is prudent to assume that it represents some type of occupant safety system. There are many air bags located in areas that are not labeled and are impossible to identify, even with careful examination. While this may at first seem to complicate the process of extrication, it may actually make operations easier. Based on the fact that it's impossible to prove the presence or absence of any particular type of air bag, the only prudent thing to do is to treat every vehicle as if it has every type of air bag. By assuming every crashed vehicle is equipped with all types of air bags, crews are more likely to develop a better routine for minimizing the dangers associated with air bags than if each vehicle had to be the subject of an extensive air bag search and evaluation.

As a general rule, to minimize the hazards of the unlikely deployment of air bags, crew members should try to stay away from the typical locations used for occupant safety systems. This means the crew member working inside the vehicle should be positioned in the center of the rear seat if possible. Of course, that won't be possible in many crashes, but it is one approach to reducing the crew member's exposure to the rare possibility of an unintentional air bag deployment.

Fig. 5–7 One example of an inflatable curtain cylinder location

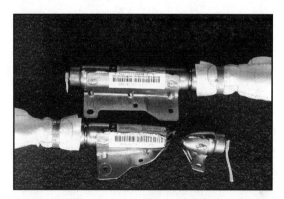

Fig. 5–8 The lower air curtain cylinder was inadvertently cut during a roof removal. While there were no problems caused by this action, compressed cylinders should be avoided when cutting.

TECHNOLOGY ADVANCEMENTS

An important part of operating safely around occupant safety systems is to stay abreast of advances in equipment design and function. Every year manufacturers make changes to their vehicles, making it almost impossible to have knowledge of how any specific system functions. When mechanics are asked about the location or technical details of a particular component in a particular vehicle, most turn to their computers and shop manuals for accurate information. If the mechanics who work on the vehicles forty hours a week can't memorize the details of the systems, it would be unrealistic to expect firefighters to do any better. By frequently researching the subject, a good general knowledge base can be developed that can be used at the crash scene to identify potential areas of concern and the appropriate actions needed to minimize the danger they pose to crew members.

Gas Lifters

While not part of an occupant protection system, another area of concern when carrying out roof procedures is the presence of gas lifters, which are used to raise hatchbacks and rear windshield glass. Again, avoid cutting any compressed gas cylinder in order to prevent the uncontrolled release of its contents and possible launching of the severed pieces. There are a few steps that can be taken to reduce the problems associated with these lifters.

Fig. 5–9 & 5–10 The presence of this lifter was not recognized until after it was cut and a mist enveloped the tool operator's upper body.

1. Locate the lifters. If the vehicle has rear windshield glass or a hatchback, there has to be some type of device to help overcome the weight of the part being lifted. A few vehicles have springs mounted in the roof to help lift, some use motors, but most of the time gas lifters are used. The key is to locate them.

2. If possible, remove the lifters. Lifting the glass or hatchback makes the lifter accessible and easier to remove. The lifter will be held in place either by a ball and socket fitting or will be bolted in place. If the ball and socket is used, a Halligan can be used to pop the lifter off of the ball. If it won't pop off and the lifter is fully extended, the piston (silver part) can be cut to disconnect the lifter from the attachment point. The cylinder (black part) shouldn't be cut because it is the part containing the compressed gas. If the lifter is bolted in place, it could be unbolted, but that probably would be a time-consuming process, and not worth the time. Again, the piston can be cut if fully extended. Even if not fully extended, when cut, the piston will extend out of the cylinder slowly.

3. If the lifters can't be removed, avoid them. Cutting above or below the lifters will eliminate most of the problem associated with the lifter. When cutting a hatchback, the tool operator should understand that the weight of the hatchback will be reduced possibly causing the remaining portion of the hatchback to spring up suddenly.

On occasion, these lifters are cut during extrications because they were inaccessible, overlooked, or were considered a reasonable risk when compared to the potential benefits.

Fortunately, when a lifter is severed it's usually held in place at both ends and located in an area surrounded by metal. If a lifter is cut, a mist may be produced that can envelope the upper body of the tool operator. While this may seem minor, avoid possible injuries by not cutting any compressed gas cylinder.

Selecting the Best Tools for the Job

Tool selection may be simple if the units on the scene only carry hydraulic tools, and there are no other units available with other tools. The officer will have a limited choice of procedures and will have to pick the best one. If there are a variety of tools available, the officer can select the tool system that is best suited for the situation based on the skills and abilities of the crew, the type of vehicle, and the type of damage. If there are enough crew members on the scene and there are two tool systems that will work well, the officer can put both tools into service simultaneously.

Hydraulic cutters are a good choice when there's a lot of deformity to small roof posts that are easily enclosed within the blades of a hydraulic cutter. If a large, badly deformed "C" post needs to be cut, the air chisel is a better choice because it isn't limited to the 6 or 7 in. of cutting depth common with hydraulic cutters. To be effective, however, the air chisel must be in the hands of a skilled operator. When a "C" post is large and in relatively good condition, a reciprocating saw can be used to quickly cut the post in one continuous cut.

A reciprocating saw may not be a good choice when the "C" post is deformed because it's difficult to determine which parts of the post are in tension and which are in compression. If the saw cuts through an area in compression, the kerf (the gap left after the metal is cut) will close behind the blade as it cuts and can cause the blade to bind. If the metal is in tension, or being pulled apart, the kerf will open up as the blade cuts, reducing the chance of a bind. A contractor-grade hacksaw will cut most "A" posts in about a minute and can make a relief cut in about the same length of time. Like the reciprocating saw, the blade is likely to bind when cutting through a post that's bent and in compression. There is, of course, no reason to use a hacksaw when power tools are available, but it's nice to know that it can do the job if one of those one-in-a-million situations comes along.

SELECTING THE BEST PROCEDURE

To select the right procedure it's important to have a clear idea of the desired results. While that may sound obvious, it's easy to start down the wrong path, only to find out that when completed, no real progress will have been made to free the victim. For example, if the goal is to access a rear seat passenger, flapping the roof to the middle, or "B" post, won't help much. It would be better to remove the roof completely to open up the back seat area. On the other hand, if a trapped victim is on the verge of respiratory arrest, and a dash roll-up is needed, the officer may elect to simply remove a single "A" post to disconnect the roof from the forward hinge pillar. While not providing a lot of space inside the vehicle, the removal of the "A" post will allow the dash roll-up procedure to be performed along with the quick extrication of the victim.

When a vehicle is resting on its side, its stability will have a big impact on which procedure is selected to free the victims. If the roof posts are contributing to the stability of the vehicle and are preventing it from rolling over, it's wise to leave the posts intact and to focus on cutting away the flat sheet metal of the roof to access the victims. That is a procedure best performed with an air chisel or reciprocating saw. In experienced hands, both tools perform equally well. If, on the other hand, the roof posts have no impact on the stability of the vehicle, a total roof removal may work, provided there is good access to all of the roof posts.

Whether the vehicle is on all four wheels or on its side, roof deformity can make flapping a roof difficult. Even roofs in good condition may need some help to get the bend started. For decades, when roof flaps were taught, students were taught that the roof should be struck with an ax to help create the bend needed for the flap. Then a movement started that taught that striking the roof was always unacceptable, and that a pike pole should be used to create the downward pressure needed to flap a roof. Unfortunately, fiberglass pike poles bend, making it difficult to apply any real downward pressure on a roof made up of bent metal. Procedure 9 demonstrates why roofs need a little help when being flapped, and one old-fashioned way to flap a roof effectively.

SUMMARY

Before beginning any course of action, the problem at hand must be clearly identified in order to develop useful options. Once the viable options are identified, the vehicle and available tools are evaluated to find a good fit between task, tools, and vehicle. When dealing with air bag systems, the application of universal precautions can reduce the chance of unpleasant surprises and crew injuries. By making the best use of the available options, the roof can be eliminated as a cause of entrapment or hindrance in the performance of other procedures.

Procedures

Procedure 1
Quickly Establishing the Presence or Absence of Inflatable Curtains to Determine the Likelihood of the Presence of Compressed Gas Cylinders

To determine if inflatable curtains are present, the driver's headliner and trim is pulled down.

Pulling the headliner down is made easier if the weatherstripping is pulled off first.

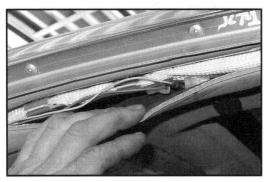

As the headliner is pulled down, the space where the air curtain would be is exposed.

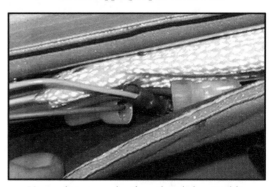

Upon close examination, the air bag and its wiring is visible.

When this headliner is pulled down, no air curtain was found.

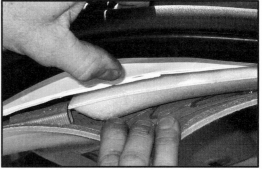

The foam in this headliner can be mistaken for an air bag if not observed carefully.

Roof Procedures

Procedure 2
Checking "A" and "B" Posts for Compressed Gas Cylinders in Vehicles Equipped with Inflatable Curtains

To determine if inflatable curtain cylinders are present, the trim is pried away from the posts.

The hazard check is simplified when the trim is knocked free during the crash.

The holes in this "B" are small, but are large enough to see the inside the post.

If there are no cylinders seen behind the inspection holes, the cutters are put in position.

The "B" post is positioned as deep as possible in the blades of the cutter.

If mulitple cuts are required, the cutters are lined up directly over the previous cut.

Vehicle Extrication: A Practical Guide

Procedure 3
Checking the "C" Posts for Compressed Gas Cylinders in Vehicles Equipped with Inflatable Curtains

The best access to the rearmost roof post is through the rear windshield.

After initially being pried away with a Halligan, the trim can be pulled by hand.

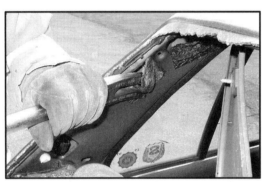

Adhesive and tapes can be removed in the search for inspection holes.

With the trim out of the way, a flashlight can be used to inspect the "C" post interior.

These inpection holes provide two choices for cutter positioning.

The lack of cylinders is apparent when the post interior is lit.

Roof Procedures

Procedure 4
Checking the Roof Edge for Compressed Gas Cylinders
in Vehicles Equipped with Inflatable Curtains

The claw of a Halligan or a screw driver is used to pry away the plastic trim above the window.

When pried away, this horizontal trim piece will seperate from the vertical trim piece.

Removal of noise dampening tape or foam may reveal examination holes in the roof structure.

If no cylinders are found in the roof structure, the cutters are positioned over the hole.

Without pulling the trim, the tool operator would be cutting blindly into the structure.

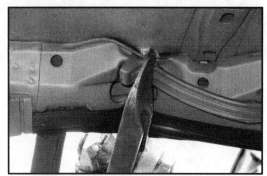
The cutter blades cut directly through the inspection hole and the clear area behind it.

Procedure 5
Lifting a Crushed Roof with a Hi-Lift Jack

A Hi-Lift jack can be used to lift a flattened roof a considerable distance. The lift will be limited to the edges of the roof because the jack will be spreading in the window openings. While the base plate of the jack is on the ground, the operator's hand or edge of a boot raises the jack's reversing lever (operating lever) upward, into the "Up" position. While holding the jack handle firmly against the steel standard (bar), the entire jack is lifted into position at the bottom of the open window.

The handle is grasped in the middle and unclipped from the standard. While holding the standard in the desired position with one hand, the operator lifts the handle upward, causing the lifting mechanism to slide up the standard. The handle is lifted until the nose is in the desired position.

The handle is grasped at the end for maximum mechanical advantage. If the operator can't reach the end of the handle, the handle should be grasped as close to the end as possible. If the handle isn't grasped at the end there will be a decrease in mechanical advantage and lifting power. The jack is operated until the roof is lifted as far as needed. It may be possible that the jack operator can develop enough lifting force that the door may come open. The tool operator should be aware of this and position in as safe a spot as possible.

Roof Procedures

Procedure 6
Lifting a Crushed Roof with a Hydraulic Spreader

Hydraulic spreaders can be used to lift the edges of a roof upward in a procedure identical to the vertical spread procedure that is used to force a door open or to create a purchase point for hydraulic spreaders. The procedure is the same, but the results are different because of the difference in the structural strength of components. In a vertical spread procedure, the roof is strong and the door is weak, so the door moves outward. When using a spreader to lift a roof, the roof is weak (damaged) and the door may be strong, forcing the roof up.

The tips are positioned so the top tip is against the edge of the roof and the bottom tip is against the door. If possible, the spreader is positioned to create a vertical lift instead of an angled lift. If lifting vertically, the door is less likely to pop open.

The tool operator moves to a safe position in case the door pops open instead of the roof moving upward. After the first lift is completed, the spreaders can be repositioned to lift another area of the roof.

Vehicle Extrication: A Practical Guide

Procedure 7
Lifting a Crushed Roof with a Hydraulic Ram

The center of a roof can be pushed upward by positioning a hydraulic ram between the inside of the roof and the floor. The tool operator should size up the vertical space available, and then select the appropriate ram. The ram is slid into position through one of the windows.

The ram is positioned in the desired area. If possible, a piece of cribbing is placed beneath the lower end of the ram.

The operating lever is positioned so the operator can reach it.

Roof Procedures

Procedure 7 *(continued)*
Lifting a Crushed Roof with a Hydraulic Ram

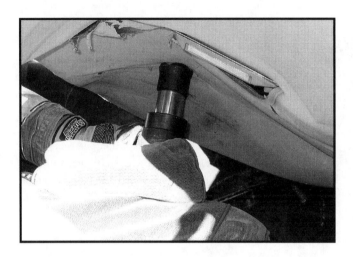

It may be possible to place a piece of 2X4 cribbing between the roof and the upper end of the ram but only with difficulty. This step will provide limited good results because of the likelihood of the upper cribbing falling out of position once the ram head starts to move.

The tool operator monitors the roof sheet metal for any sign that the metal has torn ending the effective lift.

With some effort, the ram can be repositioned to lift other parts of the roof.

225

Procedure 8
Removing Laminated Glass

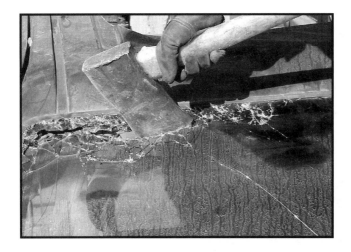

If a roof is to be flapped or removed, it is best to remove all the windshield glass to avoid having it fall on victims during the flapping or removal process. Some instructors recommend that the glass be cut along the bottom edge only and then flapped back with the roof. This may work when the vehicle is in pristine condition but won't work as well following a severe frontal collision. For a safer operation, a flat head ax can be used to tear through the glass and flexible laminate with short, choppy strokes.

A smaller tool, the Pry Ax is used in the same manner. Its shorter handle and lighter head make it easier to control than a standard flat head ax.

After the glass is struck and the head penetrates the glass, a chopping motion can be used or the serrated edge on the bottom to the head can be used to tear through the glass.

Roof Procedures

Procedure 8 *(continued)*
Removing Laminated Glass

Specialty tools designed for removing laminated glass have refined the removal procedure. To remove the windshield, the glass is struck at the top center to create a starter hole for the tool saw blade.

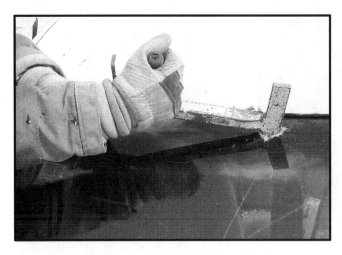

After the tool penetrates the windshield, it is removed.

The saw end of the tool is inserted in the starter hole. The tool operator pulls back and up on the tool to tear through the windshield laminate.

Procedure 8 *(continued)*
Removing Laminated Glass

As the saw reaches the "A" post, a cutting turn is made or the saw blade can be removed and reinserted after repositioning.

Another starter hole is made in the bottom of the windshield, a few inches above the lowest part of the glass.

If the saw blade is inserted at the very bottom of the windshield, the operator will have a difficult time operating the tools since it will being striking the dashboard. For this reason, it is best to start a few inches above the bottom of the windshield.

Roof Procedures

Procedure 8 *(continued)*
Removing Laminated Glass

When using any glass removal tool, the operator should wear appropriate PPE and avoid inhaling the glass dust by mouth or nose.

If working alone, one crewmember can fold the glass over to the opposite side of the vehicle. The cutting procedure is repeated on the opposite side.

With the windshield cut free of the framework, it is placed under the vehicle or in an area outside the work area. A piece of windshield glass on asphalt or concrete is extremely slippery and presents a major trip/slip hazard to crewmembers.

Procedure 9
Basic Roof Flapping Procedure
Why Striking the Roof is Often Necessary

After the cuts are made, each crewmember lifts on an "A" post until slight resistance is felt.

As the "A" posts are lifted, the curve of the roof becomes more pronounced.

To help the metal bend between the relief cuts, the roof is struck lightly with a heavy tool.

After being struck, the bow in the roof reverses, creating a crease between the relief cuts.

With the crease created, the roof folds back easily.

If traffic or weather cause gusty conditions, the roof should be tied down.

Roof Procedures

Procedure 10
Flapping a Roof with Hand Tools

Using hand tools to flap a roof would not be most firefighters' first choice of the way to go, but there are times when only hand tools are available. Remote locations, disasters and primary tool failures may leave hand tools as the only option. Flapping a roof with hand tools is not as difficult as it sounds, and in fact, it can be accomplished rather quickly with a good hacksaw, fresh quality blades, and a prying tool. The first step is to remove any trim pieces that can be removed to avoid having to cut them.

To cut the "A" post, the hacksaw is positioned near the center of the post. The center of the post is used to avoid the multiple layers of metal that may be present at the upper or lower ends of the post. In this case, the obstruction caused by the remaining lower half of the post is offset by the lack of complications encountered while cutting the post by hand.

In most vehicles, the post cut can be completed in a minute or two. Vehicles equipped with enhanced rollover protection may have posts that may make the use of a hacksaw impractical. The presence of a formidable post of this nature will become apparent if the hacksaw fails to make any progress. Once the post is cut, the blade tensioner will need to be released to remove the blade that is within the windshield frame.

Procedure 10 *(continued)*
Flapping a Roof with Hand Tools

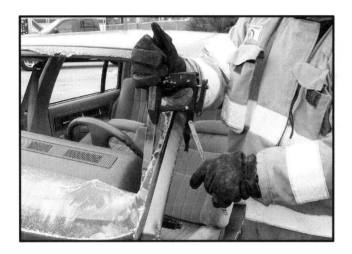

Once the blade tensioner is released, the blade can be removed from the saw, freeing it from the "A" post. The blade is then reinstalled on the saw.

The location of the relief cut is selected after checking for compressed gas cylinders. Any trim that can be removed is pried away with pliers-type tools.

Or it can be pried away with a Pry Ax or Halligan.

Roof Procedures

Procedure 10 *(continued)*
Flapping a Roof with Hand Tools

After confirming that there is not a shoulder harness assembly on the interior side of the roof, the hack saw is positioned to start the cut.

Once the saw cuts through the edge of the roof, it can be repositioned to maximize the depth of the cut. The deeper the cut is made into the roof, the weaker the roof structure will be.

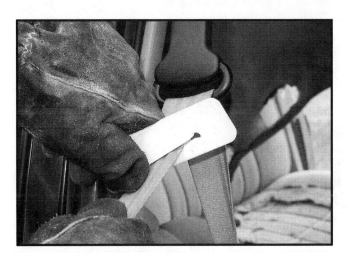

The procedure is repeated on the opposite side of the vehicle, any belts that would prevent the roof from being lifted are cut, and the roof is flapped back.

Procedure 11
Flapping or Removing a Roof with an Air Chisel

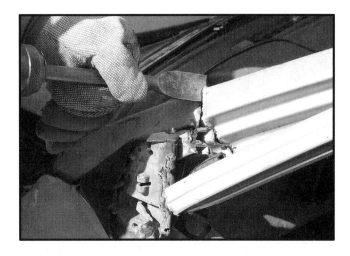

Flapping a roof with an air chisel is a good choice when severely bent roof posts would cause a reciprocating saw to bind. When cutting the smaller "A" and "B" posts, the outer perimeter of the post should be cut first using half of the blade to avoid binding. Cuts are made high enough on the post to avoid the dashboard and side mirror assembly.

After the perimeter of the post is cut all the way around the post, the core of the post can be cut with little risk of blade binding and getting stuck.

At some point during the procedure, the belts are cut. If the belts are not cut, the roof will be held in place when attempts are made to remove it.

Roof Procedures

Procedure 11 *(continued)*
Flapping or Removing a Roof with an Air Chisel

Like the "A" post, the window frame and the "B" post are first cut around the perimeter of the post to avoid having the chisel become stuck.

A systematic approach to the cuts will help the tool operator keep track of what has been cut and what has not been cut.

All sides of the post that can be seen from outside the vehicle are cut. By making the cuts an inch or two above the bottom of the window, the crewmember can avoid areas of overlapping metal, simplifying the cuts.

Procedure 11 *(continued)*
Flapping or Removing a Roof with an Air Chisel

To finish the perimeter cuts on the post, the tool operator will have to be in position to see the cuts.

After the perimeter has been cut, the core of the post can be cut, completely severing the post. To confirm that the post has been cut, the crewmember on the tool can pull outward on the post. If the areas above and below the post move out of alignment, the post has been cut. If there is no movement, the cut should be reexamined for pieces of metal that remain uncut.

To help fold the roof rearward, a relief cut is made in the roof. The purpose of the relief cut is to remove all the structural strength built into the roof edge. The edge of the roof is made of the same sheet metal as the rest of the roof, but it is shaped to create strength. By cutting through these shapes, the roof is reduced to only a flat sheet of sheet metal.

Roof Procedures

Procedure 11 *(continued)*
Flapping or Removing a Roof with an Air Chisel

As with the roof posts, the perimeter of the roof edge is cut first to avoid having the chisel bit becoming stuck when the metal binds.

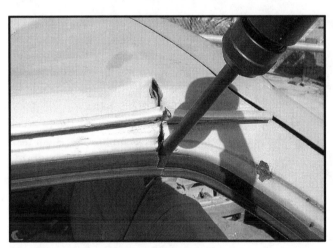

After the perimeter cuts are made, the inner part of the roof edge can be cut.

With the roof posts, belts, and relief cuts completed, the roof can be folded rearward. If the roof cannot be brought all the way to the rear of the vehicle, the roof will need to be carefully tied back to prevent it from falling into the occupant compartment during the extrication. Under gusty conditions or any other condition that may cause the roof to be propelled upward and forward, the roof should be tied down.

Procedure 11 *(continued)*
Flapping or Removing a Roof with an Air Chisel

When deciding whether to flap or remove a roof, the size of the rearmost roof post is often the deciding factor. If the roof posts are small, it may be best to completely remove the roof instead of flapping it. If the decision is made to remove the roof instead of flapping it, the forward and rear posts should be cut first, leaving the center posts to support the roof until it is ready to be removed. The cuts are made by cutting around the perimeter first, as is the procedure for cutting the other posts.

If lifters are encountered in vehicles equipped with liftbacks or hinged rear windows, the cuts should be made above or below the lifters. If possible, the door or glass operated by the lifters can be opened and the lifters pried off before cutting. This step would provide a better view of the metal to be cut as well as less of it, and a hazard is eliminated.

As soon as the liftback is cut, its overall weight will be decreased greatly and the lifters will operate, pushing the remaining structural member outward. The air chisel operator should be positioned out of the way of any part of the liftback subject to sudden movement after being cut.

Roof Procedures

Procedure 11 *(continued)*
Flapping or Removing a Roof with an Air Chisel

In some instances, even when the roof post is large, it may be necessary to remove the roof instead of flapping it. If the roof is flapped on a two-door vehicle like the one in the photograph, access to the rear seat occupants would be restricted.

To cut a large "C" post, the outer layer of sheet metal is removed to expose the inner metal and supporting structure. While basically flat, the surface of this "C" post may have too much damage for a dual or T cutter, making the curved cutter a better choice. To get started, air chisel is used to cut around the outer surface of the "C" post to create a large inspection or access hole.

The area within an inch or so of the window is avoided because there will be more structure in that area to support the glass. Cutting too close to the window will slow the entire procedure.

Vehicle Extrication: A Practical Guide

Procedure 11 *(continued)*
Flapping or Removing a Roof with an Air Chisel

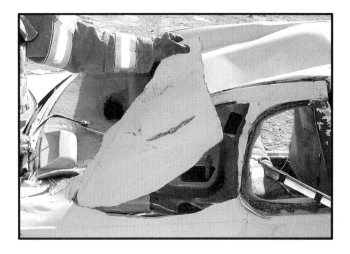

When the large piece of sheet metal is removed, the tool operator will have a good view of the structure that will need to be cut to sever the "C" post. With the outer layer removed, large holes in the inner layer of sheet metal are visible.

A quick size-up is made to determine the quickest and easiest way to sever the remaining metal. Time and air can be saved by utilizing the holes that are present by simply *connecting the dots.* After all the dots are connected and the final cuts around the perimeter are completed, the procedure is repeated on the other side.

Any remaining belts that would hold the roof on are cut, and the roof is ready to come off.

Roof Procedures

Procedure 12
Flapping a Roof with Hydraulic Tools

Flapping a roof with hydraulic tools is a procedure performed almost as often as popping a door. The basic roof flap procedure requires the severing to both "A" posts and two relief cuts. This procedure is quicker than the roof removal procedure and may be a good choice for large vehicles like station wagons that have many roof posts.

The "A" posts are cut first using the cutters. The cutters are positioned low and perpendicular to the post with the post positioned deep in the cutters. By positioning the cutter perpendicular (at a 90-degree angle) to the post, the smallest amount of metal will be cut. By cutting less metal, the cutters have a greater chance of completing the cut in one attempt.

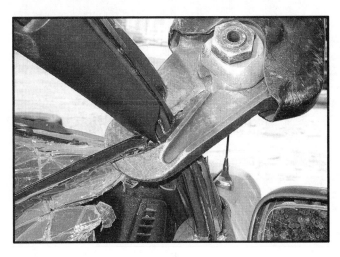

As the cutters are closed, the post may slip along the blades if the teeth do not grasp the metal. Except for straight blade cutters, most cutters are curved inward at the end of the blades to hold the metal in the jaws of the tool.

Vehicle Extrication: A Practical Guide

Procedure 12 *(continued)*
Flapping a Roof with Hydraulic Tools

Relief cuts are made to remove the strength and integrity achieved at the edges of the roof. If access is needed only to the front seat area, the relief cut can be made in front of the "B" post. If access to the rear seat area is required, the "B" posts are cut low and relief cuts are made in front of the "C" posts.

A few inches can make a big difference, so the cutter is positioned as close to the "B" post as possible with the blades open fully. A little up and down adjustment of the cutter position will help maximize the depth of the cut.

Before making the cut, the tool operator should make sure the cutters are clear of any shoulder harness brackets that would make the cut more difficult. The shapes of the roof edge are what give the roof resistance to bending and can be seen in the photograph.

Procedure 12 *(continued)*
Flapping a Roof with Hydraulic Tools

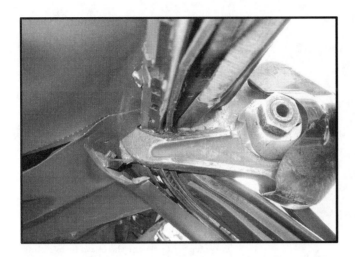

When the relief cut is made, a distinctive snapping of metal sound is often heard. This sound is an indication that the cut has been completed.

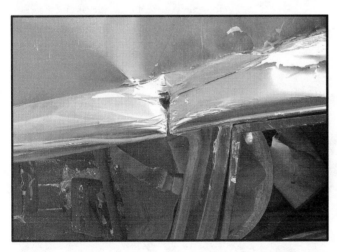

When the cutters are removed, the cut can be examined to verify that the structure was completely severed. After the structure at the edge of the roof is severed, the only resistance to flapping the roof will be the flat sheet metal of the roof.

If there is little deformity to the sheet metal that lies between the two relief cuts, the roof can be folded back by hand. If there is significant deformity, a come-along may be needed to overcome the folds and bends in the roof. Other options include extending the length of the relief cuts to further weaken the structure or to cut the sheet metal between the relief cuts with a reciprocating saw or air chisel. By combining two or more tool systems, the roof flap can be converted into a partial roof removal.

Vehicle Extrication: A Practical Guide

Procedure 13
Flapping a Roof with a Come-along

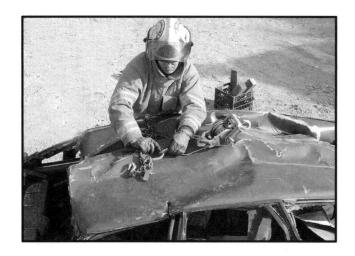

Damage and deformity to a roof can make it difficult to flap back manually after all the cuts have been made. The folds and bends created in the flat sheet metal sections of the roof will stiffen the roof structure making it difficult to fold even with relief cuts. To defeat this resistance, the power of the come-along can be used to fold the roof back. While time consuming, this procedure may be necessary when the roof is too heavy to remove or when the proper removal tools aren't available. To start, the come-along is set on the roof and is attached to the master link of a chain set. This step will prevent the chain from falling into the vehicle during the rigging process.

The end of the chain is brought around the "A" post on one side of the vehicle then passed to the other side.

The second "A" post is wrapped in a mirror or reverse image of the first post. The end of the chain is then brought back and attached to the chain shortener.

Roof Procedures

Procedure 13 *(continued)*
Flapping a Roof with a Come-along

After the roof chain has been shortened to the desired length, the come-along is moved to the side of the vehicle so it can be walked back to the rear of the vehicle.

While the roof is being rigged, a crewmember rigs an anchor chain at the rear of the vehicle. When positioning the master link, an effort is made to rig the chain so the come-along will operate in as much of a horizontal position as possible. The ideal placement on this station wagon would be on the roof, but it would be very difficult to operate the come-along from that location. The next best choice is a position in the rear glass area. This position, while not perfect, is better than having the come-along pointed skyward or being on the roof. For sedans, the trunk lid is the best position.

The come-along is walked back and connected to the anchor chain master link.

Vehicle Extrication: A Practical Guide

Procedure 13 *(continued)*
Flapping a Roof with a Come-along

To prevent crushing, cribbing is placed between the body of the vehicle and the anchor chain. The goal of preventing crush is not cosmetic but is aimed at making each stroke of the come-along handle more effective. As the handle is operated, cable is wound onto the drum, which in turn pulls the roof up. If the operation of the come-along causes the metal to crush instead of pulling the roof up, time and effort would be wasted.

It is always best to run come-along cables so they are not damaged by contact with other objects. In this procedure, that is nearly impossible. To minimize any chance of damage to the cables, floor mats, cribbing or other objects can be positioned to protect the cables.

Pulling directly rearward could cause the roof to actually move downward instead upward as is needed. To get the proper angle or lift for the pull, a piece of cribbing is placed under the master link. The cribbing should be on or just behind the relief cuts that will form a crease and bend across the roof.

Roof Procedures

Procedure 13 *(continued)*
Flapping a Roof with a Come-along

With the cribbing in place, the officer calls for the slack to be taken out of the system, and the rigging to be checked. The come-along is then ready to pull the roof up. To help the roof get started in the right direction and to counteract any movement downward, one or both of the "A" post are lifted upward by crewmembers. If the roof is badly damaged and deformed, development of the crease may be helped by striking the desired crease location with a sledge hammer or back of a flat head ax. While some may discourage this step, on occasion, it is exactly what is needed to successfully flap the roof.

Once the roof starts up, the come-along will become easier to operate.

The come-along can be operated until the roof is pulled completely back or adequate space and access is obtained. To avoid having somebody walk into the handle of the come-along after the procedure is completed, the handle is removed but left in the area of the come-along.

Procedure 14
Removing a Roof with an Electric Reciprocating Saw

After a rollover crash or severe side impact, a roof may be so badly deformed that flapping it may not be an option. The deep creasing and folds will prevent any attempts to fold it rearward. In these situations, removal is a good option.

The "A" post or the rearmost roof posts are cut first leaving the center posts intact to support the weight of the roof until the last cuts are made. The saw blade should be positioned to cut through the least amount of metal possible. To avoid the dashboard and any reinforcement, the cut can start a few inches above the bottom of the "A" post.

Vehicles equipped with hatchbacks will have lifters that are used to help lift the hatchback as it is opened. These devices should be avoided to reduce the amount of metal that needs to be cut and sudden release of pressure if a cylinder was to be breached.

Roof Procedures

Procedure 14 *(continued)*
Removing a Roof with an Electric Reciprocating Saw

It may be possible to see the lifters and position the saw to cut above or below them. No matter how the post is cut, the saw operator should not be positioned over the hatchback door as it is cut. When the part of the door is cut away, the weight of the door will be reduced, changing the original weight balance the door had before it was cut. This change in weight will result in some part of the door being pushed outward by the lifter.

If possible, the lifter should be exposed and removed to simplify the operation.

Some time during the procedure, all belts connected to the roof posts and above should be cut away.

Vehicle Extrication: A Practical Guide

Procedure 14 *(continued)*
Removing a Roof with an Electric Reciprocating Saw

After the front and rear posts are cut, the remaining center posts are cut. By leaving these posts for last, it may not be necessary to tie up personnel for as long with the task of supporting the roof.

The stresses placed on the roof posts may be difficult to recognize when starting the cuts. If the saw starts to bind and slow down during the cut, it is probably caused by the compression on the roof post that is causing the kerf to close behind and around the blade. Before the blade binds too badly, the saw can be removed from the cut and repositioned on the other side of the post. The cuts are lined up and completed. If the blade starts to bind again, the saw can be repositioned again. Instead of approaching from the front or rear of the post, the cut can be started from the inside or outside of the post.

With the roof completely removed, crews have more options for access, treatment, and removal of the victims.

Roof Procedures

Procedure 15
Removing a Roof with Hydraulic Tools

The use of hydraulic tools is one of the simplest yet most powerful solutions to the need for a roof removal. Hydraulic cutters, in good working condition, are capable of powering through metal that would cause other tools to bind and stop operating. A great deal of finesse isn't needed to complete the task, making it a good all-around choice of tactics. To remove a roof, the cutters are positioned low on the "A" post, but not so low as to involve the dashboard in the cut. When cutting a post, most roofs will produce a distinctive *thump* as the post is severed.

Leaving the "B" post for last, the trim area inside the rearmost post is removed. The quickest way to expose this area is to pry it open with a Halligan or other prying tool.

With the interior trim out of the way, the crewmember can observe, locate any cylinders used for the vehicle's air bag system. If cylinders are present, making the roof post cut undesirable, other options should be considered.

Procedure 15 *(continued)*
Removing a Roof with Hydraulic Tools

With no cylinders present, the cut can be made. To minimize the number of cuts needed to sever the post, the cutters should be positioned across the shortest distance from the front of the post to the rear of the post. This could be high, low, or in the middle of the post.

Any belts that would hold the roof on are cut. The "B" post is cut after locating the shoulder harness bracket and avoiding it. The backing place in this assembly is heavy metal, making it an object to avoid when cutting the post. More than one set of cutter blades have been fractured when the tool operator did not locate the assembly and tried to pierce it with the tips of the cutters. Cutter tips ordinarily are not capable of piercing substantial steel and should not be used in that manner.

A coordinated effort is used to remove the roof, especially the large ones. A full-size or luxury sedan roof will require four crewmembers to safely carry it out of the way.

Roof Procedures

Procedure 15 *(continued)*
Removing a Roof with Hydraulic Tools

If the decision is made to remove the roof on a vehicle with a very large "C" post, the crewmembers charged with cutting the post will be in for a challenge. There are a couple of approaches that can be used to cut the large post.

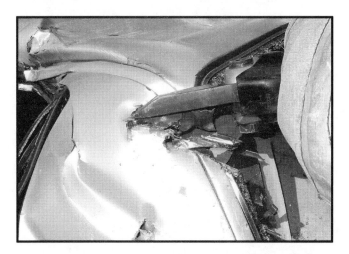

The first procedure utilizes a series of pie cuts to reduce the width of the post. To start the post removal, an angled cut is made in the "C" post.

The next cut is made by positioning the cutters so the tips are in the same location as the first cut. The rest of the cutter blades are positioned to create the pie cut or wedge.

Procedure 15 *(continued)*
Removing a Roof with Hydraulic Tools

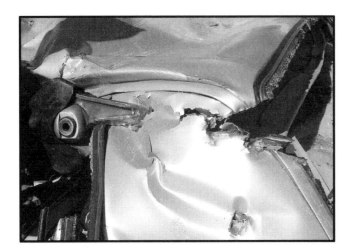

With skill and good luck, the wedge will fall out of the "C" post as the second cut is completed. If the wedge fails to fall out, the metal holding the wedge in place is identified and cut.

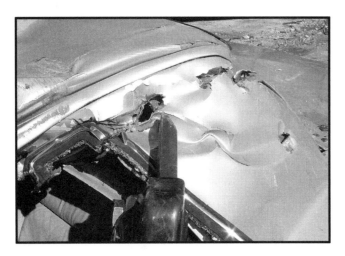

The procedure is repeated on the other side of the post, starting with one cut, then making the second to complete another wedge cut.

If the wedge remains connected in the cut, it can be pulled outward and cut.

Roof Procedures

Procedure 15 *(continued)*
Removing a Roof with Hydraulic Tools

With two pie or wedge cuts completed, the cutters can be brought into position to finish the cut. Cutters with long blades are better suited for this procedure as they may be able to finish the procedure by reaching from one pie cut to the other. If the cutter blades won't open wide enough, additional cuts from the front or rear of the post will be needed.

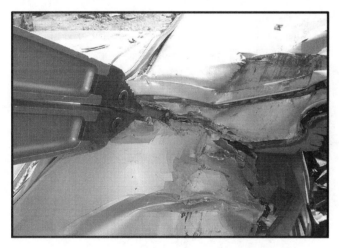

Another solution for situations when the cutter blades are small is to use the spreaders to tear through the remaining metal. The spreader tips are inserted into the pie cut as close to the center of the post as possible.

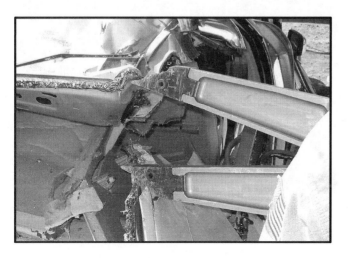

The spreader arms are opened, tearing the remaining metal. This procedure works because the pie cuts are being used as relief cuts. Relief cuts are made to start a tear that will continue when spreading force is applied to the metal.

Vehicle Extrication: A Practical Guide

Procedure 16
Making a Three-sided Roof Cut with an Air Chisel

Air chisels and reciprocating saws are both good choices when a three-sided roof cut is necessary to free victims. If the roof has a lot of deformity caused by a rollover or severe side impact, the air chisel may be a better choice over the reciprocating saw because it is less likely to bind while cutting through twisted and bent metal. When cribbing the vehicle to stabilize it, cribbing can be placed in any location, including the top of the roof posts. In this procedure, the top of the roof posts and drip edge can be used because the framework of the roof will remain intact after the cuts are completed.

The cut is started in one of the upper corners, a few inches from each edge. By staying away from the edges of the roof, reinforced areas of the roof can be avoided. Typically a single cut is made from the upper corner that proceeds downward, but the cut can be just as effective if started in the middle and worked in one direction then the other.

While the T and dual cutter work well on smooth sheet metal, the curved chisel bit is the best choice when the roof is badly deformed. Invariably, the change in the shape of the metal causes the T and the dual bits to punch all the way through the metal, making the cutting process tedious and nonproductive. When using the curved cutter, the outer half of the blade should be kept outside the cut when possible.

Roof Procedures

Procedure 16 *(continued)*
Making a Three-sided Roof Cut with an Air Chisel

The downward cut is made to a point a couple inches above the lower edge of the roof. By stopping at this point, the reinforced edge of the roof is avoided and there is less of a chance of hitting an undetected air curtain cylinder. When the roof metal is folded down, the end of the vertical cuts will help form the fold line.

The second cut extends from one upper corner of the roof to the other. Either end of the roof can be used as a starting point. The ease with which the cut is made is based in great part on the practice and skill of the operator.

A second vertical cut is made similar to the way the first vertical cut was made. The photograph shows why a curved chisel bit is preferred when cutting sheet metal that is deformed. If a T cutter were used on this roof, the operator would have a difficult time keeping the bit from punching through the roof.

Vehicle Extrication: A Practical Guide

Procedure 16 *(continued)*
Making a Three-sided Roof Cut with an Air Chisel

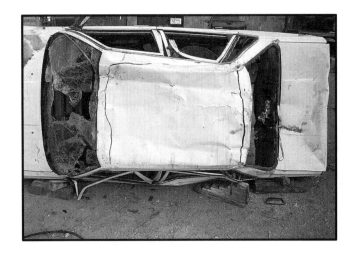

All three cuts are completed and the roof is ready to be pulled out.

The kerf created by an air chisel bit is not very wide and it may be necessary to pry the roof metal out with a Halligan.

The sheet metal is folded downward with a smooth motion to avoid rocking the vehicle. To assist in the creation of the fold or crease, a crewmember can step on the metal near the fold as the flap is being pulled out and down.

Roof Procedures

Procedure 16 *(continued)*
Making a Three-sided Roof Cut with an Air Chisel

The headliner will usually be made of a cloth material, cardboard, or a foam board product. Often, the headliner can be removed by hand while other cases may require a knife or scissors to cut through the material. If a knife is used, the crewmember using it should first check behind the material to be sure all victims are clear of the area. On occasion, a second layer of sheet metal will be found beneath the outer sheet metal layer. While usually unanticipated, the second layer is removed in the same way the outer layer is removed. The cuts on the second layer should be positioned a few inches inside of the first cuts to provide the tool operator with room to maneuver the air chisel.

Much more common is the combination of an outer layer of sheet metal and a couple of metal ribs that extend from one side of the vehicle to the other. These ribs are usually lightly welded but may be difficult to remove by hand. Some roofs won't have ribs but will have metal rods that support the headliner. These rods can be removed by hand, and the headliner can be torn away.

With the rib intact, there is plenty of room to work, but the space is obstructed. Removal of the rib would improve the work area greatly.

259

Procedure 16 *(continued)*
Making a Three-sided Roof Cut with an Air Chisel

The ribs can be quickly removed with the air chisel. To reduce bouncing of the metal and to provide as large an unobstructed space as possible, the rib is cut close to the edge of the roof.

To fold the rib down, the wires that run to the interior light fixture need to be cut.

To avoid problems of the rib acting like a spear after the top is cut and folded down, the bottom joint is cut to completely remove the rib.

Roof Procedures

Procedure 16 *(continued)*
Making a Three-sided Roof Cut with an Air Chisel

With the rib removed, there is more unobstructed work area. Crewmembers who are working to treat and remove the victims may find the roof metal in their way and cumbersome to work around. The drip edge of the vehicle in the photograph is relatively high because of the way the vehicle settled when put on its side. The position of the vehicle and the height of the drip edge make the problem worse.

To remove the problem, a fourth cut can be made to completely remove the sheet metal. This cut results in more unobstructed work area, but it also creates a very sharp edge that should be covered or avoided. The weight of a victim's body pressing down on the sharp metal can cause terrible lacerations that should be guarded against.

With both the rib and the sheet metal removed, the work area is wide open and unobstructed. After performing this procedure, the cribbing should be checked to ensure that it is still supporting the vehicle properly. As the sheet metal is removed, the distribution of weight on the vehicle will change, necessitating the cribbing check and possible adjustment.

Vehicle Extrication: A Practical Guide

Procedure 17
Making a Three-sided Roof Cut with a Reciprocating Saw

A reciprocating saw is a good choice of tools when making a three-sided roof flap because the metal of the roof is fairly thin and flat. The thin, flat metal reduces the chances of blade binding, allowing the cuts to be made in a long, continuous manner. A starter hole made with a Halligan will provide a good starting point for the cut. After any required cribbing is put in place, a starter hole is made by driving the pike of a Halligan through the sheet metal.

To create a cut as large as possible, the Halligan is positioned at one of the corners of the roof a few inches from each roof edge. By positioning a couple inches from the edges, reinforcements and heavier structural members can be avoided. Before making the starter hole, the interior of the roof area should be viewed to make sure there are no victims in the area of the planned Halligan penetration. The pike of the Halligan is held in position and then struck with a sledge hammer or ax.

The Halligan is removed and the saw blade inserted. For this type of cutting, a blade about four inches in length should be long enough to cut all the metal encountered. The TPI count is a matter of personal preference, but metal cutting blades with 10 or 14 TPI are a couple of good sizes to try. To avoid bouncing while cutting, the shoe of the saw is placed directly against the metal.

Roof Procedures

Procedure 17 *(continued)*
Making a Three-sided Roof Cut with a Reciprocating Saw

A vertical cut is made extending down to the low edge of the roof. Instead of stopping a couple inches from the edge, the cut is extended downward until there is some resistance, indicating the edge of the roof structure has been met. While making the cut, some forward pressure against the saw will reduce the amount of bouncing that can occur when sheet metal is cut.

After the first vertical cut is made, the saw blade is reinserted into the starter hole and a cut is made along the high side of the roof from one end of the vehicle to the other. To make the biggest opening possible, the saw operator should be careful not to allow the cut to drift downward towards the center of the roof during the cut. The saw will move quickly across the surface of the single thickness sheet metal but will slow when a structural rib is encountered. Roof ribs are usually about 3–4 in. wide and extend from one side of the roof to the other. It may take a few seconds to cut the rib, but after cutting it, the speed of the cut will pick up.

As the cut approaches the edge of the roof, with some practice and skill, the saw operator can gently turn the saw into the next downward vertical cut instead of stopping the cut and repositioning the saw. The advantage of making the curved transition to the next cut is that there is no need to make another starter hole.

Vehicle Extrication: A Practical Guide

Procedure 17 *(continued)*
Making a Three-sided Roof Cut with a Reciprocating Saw

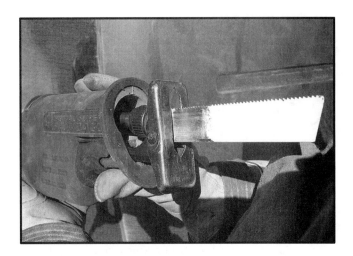

If the curved cut cannot be made and the Halligan is not available to make another starter hole, a plunge cut can be made with the reciprocating saw. A plunge cut is a cut made through any material when an edge or starter hole is not available to start the cut. A plunge cut is made by slowly operating the saw blade against the surface until the teeth at the front of the saw remove enough material that the blade penetrates the material. To obtain the smallest angle possible between the saw and the surface, the blade is removed then reinserted upside down.

This roof had a small depression in the surface that created a small edge that was used to start the cut, making reversal of the blade unnecessary. In these situations, the saw is operated slowly until the blade penetrates the metal then is brought up to speed for the rest of the cut.

After the blade penetrates the metal, the saw is rotated up to the normal operating position with a foot against the metal to reduce bouncing.

Roof Procedures

Procedure 17 *(continued)*
Making a Three-sided Roof Cut with a Reciprocating Saw

The vertical cut extends downward until the low end of the roof is approached. When the saw makes contact with the larger structural members of the roof, the cutting action will be slowed. For best results, the cut should be stopped when resistance is encountered. If the cut were to be continued, and the edge of the roof severed, it could impact the effectiveness of the cribbing. Completely severing the roof may seem like a good idea, but it can complicate the process of folding the roof down.

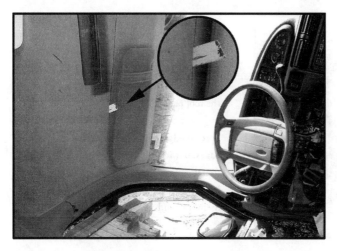

When the saw is being operated, there is a natural tendency to forget that the other end of the blade is in the passenger compartment of the vehicle. Before and during the cuts, the saw operator should confirm that the victims have not moved into the path of the blade as it enters the passenger compartment.

With all three cuts completed, the upper corners of the roof can be pulled outward to fold the roof downward. The ribs at the low part of the roof may provide a little resistance but can be overcome with manual force.

Vehicle Extrication: A Practical Guide

Procedure 17 *(continued)*
Making a Three-sided Roof Cut with a Reciprocating Saw

As this roof was folded down, the two ribs at the rear of the vehicle folded down without any problem. But two ribs at the front were sturdy enough that, instead of folding down, the sheet metal of the roof separated from them, leaving the ribs in place. For more unobstructed space, the ribs are cut away with the saw.

After the roof is folded down, the inside of the roof may be slippery and difficult to work on because of the incline. To eliminate this problem, a fourth cut can be made along the bottom edge of the roof to completely remove the roof panel. There are pros and cons to making the fourth cut. If the cut is not made, the bottom edge will be smooth and free of sharp edges, but it will be difficult to stand on. If the fourth cut is made, a jagged edge will be formed that can cause some problems for crewmembers and victims alike.

Whichever decision is made about the fourth cut, the procedure provides excellent access to every seat in the vehicle, making it a good choice for long vehicles like vans.

Roof Procedures

Procedure 18
Removing the Center of the Roof with a Reciprocating Saw

When the vehicle is not being supported at all by the roof posts, the reciprocating saw can be used to make either a three-sided cut or to remove the entire center portion of the roof, leaving the roof posts and the roof side edges intact. The beginning of the cut may take a few seconds longer than anticipated because of the reinforcement in the front and rear edges of the roof. It is best to make the lower cut the first cut to avoid the binding problem that can occur if the upper cut is made first. If the upper cut is made first, the weight will press down on the blade causing it to bind.

Firm pressure should be applied against the saw to keep it from bouncing when it enters the flat sheet metal area.

Usually one or two ribs will be encountered during the cuts. Like the front and rear edges, these cuts take a little longer as the blade makes contact with the reinforcement. Smooth, steady cutting will limit the chance of the blade binding. It is best to cut a little slower in one long cut than to try to cut fast only to repeatedly bind the blade.

Vehicle Extrication: A Practical Guide

Procedure 18 *(continued)*
Removing the Center of the Roof with a Reciprocating Saw

The procedure is repeated to make the upper cut. Bouncing of the metal may occur more often with the bottom cut completed. If necessary, another crewmember can support the roof to give it some stiffness.

As the upper cut approaches completion, it needs to be secured by another crewmember to prevent it from falling into the occupant compartment and causing injuries.

As the center portion is removed, it should be moved away from the work area to prevent the crewmembers from slipping on it.

Roof Procedures

Procedure 19
Roof Flap with Hydraulic Cutters, Vehicle on Its Side

A common procedure performed on vehicles resting on their sides is the roof flap. Unlike the standard roof flap performed on vehicles on all four wheels, this roof flap involves flapping the roof to the side instead of rearward. The upper roof posts are cut; the order is not important. In this case the "A" is cut first.

The "B" is cut after locating the shoulder harness bracket and avoiding it.

The "C" may require a couple of cuts to complete. Again, any shoulder harness brackets are located and avoided.

Procedure 19 *(continued)*
Roof Flap with Hydraulic Cutters, Vehicle on Its Side

A relief cut is made in the roof from the upper rear windshield area. The relief cut will sever the reinforcement found at the edge that would resist being bent.

With the headliner removed, this interior photograph shows the location of the reinforcement as the hydraulic cutters sever it.

Another relief cut is made at the other end of the roof. The relief cuts should be lined up so the metal will bend from one cut to the other as the roof is pulled downward.

Roof Procedures

Procedure 19 *(continued)*
Roof Flap with Hydraulic Cutters, Vehicle on Its Side

With all roof posts cut and the relief cuts made, the roof is pulled outward and downward. If the sheet metal bows outward at the location of the desired fold, it may have to be struck lightly to bend it inward. If the roof is struck, crewmembers should recognize that the occupants may be in a pile near the ground in the exact area where the roof needs to be struck. If there are occupants in that area, the roof should not be struck because of the risk of creating an injury.

While this is a common procedure, the end result is a little messy. There are multiple obstructions to trip over, and the roof may remain a little elevated, making it a difficult area to work in.

Sometimes the relief cuts create the beginning of a fold, sometimes they do not. If the sheet metal at the edge of the roof is badly damaged, the relief cuts may not help the roof fold down. In these situations, the roof posts will twist or bend when downward pressure is exerted on the roof.

6

Interior Procedures

Interior procedures used to free victims are usually carried out after access to the victim has been gained by removing the doors, sides, or roof of the vehicle. While the outside crew is handling the exterior procedures, the crew member assigned to the interior of the vehicle can evaluate any entrapment and start developing options to free the victim from the wreckage. The crew member assigned to the interior will have a good view of the entanglement and may be able to ask the victim questions about the entrapment that will help determine which procedures will be needed to complete the extrication. For example, when talking to a victim who appears to have a foot trapped in the toe pan area, the inside crew member may learn that the foot isn't actually trapped, but appears that way because the victim can't move his lower leg up because the dashboard is in the way.

During the size-up, the inside crew member will be working with the other crew members to accurately identify the parts of the vehicle that are preventing the victim from being removed from the vehicle. After identifying the source of entrapment, the degree to which the component must be moved can be determined. If the inside crew member can accurately assess the degree of entrapment, a course of action can be laid out with a minimum amount of time wasted performing unnecessary or inadequate procedures.

After getting a report from the inside crew member and then considering the available options, the officer should remember to first utilize the normal vehicle features that may free the victim without cranking up any tools. For example, if a victim appears to be trapped behind the steering wheel, there are a few simple steps that may solve the problem. A tilt, or telescoping wheel, may be operated to move the steering ring out of the way, or the seat can be moved rearward to create more space between the victim and the steering wheel. In severe frontal crashes, a steering wheel that appears to be pinning a victim may be actually only lying against the victim. If the steering column support brackets are broken loose during the crash, the steering column and wheel will simply fall downward until making contact with the occupant. What may appear to be a complicated extrication may be resolved by picking up the steering wheel with one hand to move it out of the way.

If these procedures don't appear to be the solution, the next simple step is to cut the ring with bolt cutters or hydraulic cutters. The bolt cutters may be difficult to position and operate, but they can be put in service more quickly than most hydraulic cutters. If it is obvious that not only the steering ring is trapping the victim, but the dashboard is also a problem, the dash roll-up or dash-lift procedure may be a better choice of procedures.

Relocating a steering column with a hydraulic spreader or come-along

The procedure of relocating the steering column with hydraulic spreaders or a come-along, while popular before the mid–1980s, has fallen out of favor because of its complicated set-up procedures, limited success, and perceived danger to the occupant. When performing the steering column relocation, the perceived danger to the occupants is based on the way the column is rigged.

In both procedures, the steering wheel and upper part of the column are pulled toward the front of the vehicle with a tremendous amount of pulling force as the come-along takes up cable or the spreaders close. As the top part of the steering column assembly is being pulled forward, a fulcrum point is created where the steering column is supported in the dashboard. The concern is that as the top part of the column is being pulled forward, the bottom half of the column will break loose, swing out from under the dashboard, and strike the victim. The concern is understandable because the steering column actually becomes a large Class One lever. The steering column relocation procedures have been relegated to the second tier procedures group; the procedures are good if nothing else will work but shouldn't be considered a first choice.

Some may totally dismiss the procedures as simply unsafe and inappropriate in any setting, but there is no advantage to removing a procedure from the group of options, especially in situations when nothing else will work. As soon as it's said that a procedure should never be performed, a crash comes along where the procedure turns out to be the only one that will free the victims. The key is to understand the risk and weigh it against the benefits before deciding whether or not to use any particular procedure.

Performing a dash roll-up with hydraulic tools

The dash roll-up procedure provides excellent results in severe frontal collisions and entrapments. Thanks to the development of hydraulic rams, entire dashboards, steering columns, and firewall assemblies can be pushed forward and

Interior Procedures

away from the occupant area to free trapped victims. In recent years, the rams have been improved with the introduction of optional attachments that create larger bases, and ram heads that grip the hinge pillars more securely. Another big jump in hydraulic tools was the introduction of the three-piece telescoping ram. The telescoping ram design allows one ram to push a greater distance than the traditional two-piece ram (see Figs. 6–1 & 6–2).

If hydraulic rams aren't available, a Hi-Lift jack can be used to push the dashboard forward with good results.

The dash lift is a procedure that works just as well as the dash roll-up, but doesn't require rams to perform the procedure. As the procedure name indicates, the dash lift frees victims by lifting the dash upward, in comparison to the dash roll-up procedure, which pushes the dashboard forward. Both the dash lift and dash roll rely on weakening some structures, and strengthening others for success. In the case of the dash lift, the integrity of the forward hinge pillar will contribute greatly to the success of the procedure. If the pillar and the end of the dashboard are gone, or badly damaged, the dash lift may not be the best choice because there may be nothing to push against. The dash roll procedure may be made more difficult by the same type of damage but may have better results because there will be more choices of push points.

If either the dash lift or roll is selected as the best choice of procedures at a crash scene, the victim's feet and lower legs should be examined to determine if the moving metal will make additional contact with the lower extremities, or if it will move past them without any problem. Fortunately, when the driver's feet are trapped between the pedals and the toe pan (angled floor area), the dash roll or lift will often pull the pedals up and away from the floor, freeing the driver's feet.

Figs. 6–1 & 6–2 When performing a dash roll-up, the three-piece telescoping ram can achieve a longer push than a similarly sized two-piece ram in the retracted position.

Removing pedals

There are other options for removing pedals as a source of entanglement if the dash roll or lift aren't necessary or won't work. The simplest procedures include cutting the pedal arm with a manual or engine-driven hydraulic pedal cutter, standard full-size hydraulic cutter, or a pneumatic whizzer. The manual or engine-driven hydraulic pedal cutters are the best choice because they are lightweight and small, making them easy to use in tight areas where larger tools won't fit. If there is plenty of space, the standard hydraulic cutters can be used, but the tool operator will need help from another crew member in keeping the ends of the cutter blades away from the victim's feet.

When setting up to make the cut, the tool operator will be focusing on positioning the pedal arm as deep as possible into the cutter and may not be able to see how close the tips of the cutters are to the driver's foot as they close. The second crew member can use a flashlight to illuminate the area and stop the tool operator if a problem develops. The whizzer usually isn't a good choice for most pedal cutting operations because of the sparks it produces as it cuts steel. The shower of sparks can ignite carpeting and other ordinary combustibles and can be an unacceptable risk in the presence of a fuel leak. Additionally, the whizzer may use two or three air cylinders to cut a pedal arm, an indication of the time needed to complete the procedure.

Relocating a pedal with a length of rope

If the crashed vehicle is small and lightweight, pulling the pedal out of the way with a piece of rope is another option; if the vehicle is larger and heavier, a come-along can be rigged to pull the pedal upward or to the side. These are useful options that can be employed based on the type of entanglement encountered.

Using a Hi-Lift jack to create interior space

Side-impact crashes create different entanglement problems than frontal crashes, but the entrapment problems are often eliminated during the victim access phase of the operation when the doors, roof, and side of the vehicle are removed. In some severe side-impact collisions, an interior spreading operation may be needed to free the victim; this can be accomplished with a hydraulic ram, or Hi-Lift jack (See Procedure 3). When using tools for interior spreading, careful consideration should be given to how the tool would move if the portion of the vehicle that's being spread were to suddenly break free. Sudden, unexpected tool movement could cause the tool to make contact with the victim and possibly cause an injury. Careful evaluation and planning can reduce the chance of this type of problem.

Relocating a seat

Some crashes may require a seat to be removed to gain access to a victim on the opposite side of the vehicle, in the rear seat, or under the dashboard. Generally, relocating a seat that is occupied by a victim is avoided because of the sudden movement that occurs when the seat track fails (See Procedure 3). When moving or removing a seat with tools, once the seat is broken free, it's best to continue to move the seat with the tool instead of trying to move it by hand. The seat cushions and springs that are easily overpowered by the tools can be a source of frustration when attempts are made to move them by hand.

Interior Procedures

SUMMARY

Interior procedures needed to disentangle a trapped occupant should be planned out while crew members are making access through the doors, sides, and roof of the vehicle. Careful inspection of the components responsible for the victim's entanglement will provide a basis for selection of tools and procedures to free the victim. When selecting the best tools and procedures, a mental image should be developed of how the components and tools will move and what impact that movement will have on the victim. Before committing to the selected procedures and tools, the officer should look forward one additional step to consider if, when completed, the procedures will permit the victim to be removed from the vehicle. If the procedures won't free the victim, the plan should be adjusted as necessary to allow the victim to be removed from the vehicle upon completion.

Procedures

Vehicle Extrication: A Practical Guide

Procedure 1
Interior Spread with a Come-along

This type of high speed side impact can create serious interior space problems and entanglement. While it is usually best not to move a vehicle when the victims are trapped inside, there are times when a moving vehicle may benefit the victims more than trying to work around a large obstruction. The company officer in charge of this type of crash will have to make a tough risk-benefit analysis and decide how the victims would be better served. When the floorboard folds like an accordion trapping the passenger's feet, the officer may decide to move the vehicle for better access or risk having the victim die while still entrapped.

Another option is to use a strong obstruction as an anchor and slowly pull the sides of the vehicle outwards. Before deciding to use the obstruction as an anchor point, it should be examined for strength and structural stability. If there's any sign of weakness, damage or other indication that the anchor may fail when under load, it shouldn't be used. If the anchor is rejected and it's still necessary to spread the interior, an alternate anchor point such as fire apparatus could be used instead. If the decision is made to continue with the interior spread, the victims' lap and shoulder belts should be disconnected before rigging the vehicle for the pull. A loop is then made with a chain around the "B" post and the obstruction and shortened with the chain shortener.

A chain is wrapped around the "B" post on the opposite side and attached to the block of a come-along.

Interior Procedures

Procedure 1 *(continued)*
Interior Spread with a Come-along

The come-along is walked back and attached to the anchor chain that is attached to a good, heavy anchor. The officer directs the tool operator to take the slack out of the come-along system and then has the crew check the rigging. After the rigging is checked, the come-along is operated.

The crew should monitor the expanding interior space, along with the "B" posts, pillars and the anchors and stop the operation if failure of any of those components appears imminent.

While the procedure may not create a tremendous amount of space, it may relieve some of the problems associated with the accordion-type of entrapment problems that can occur with severe side impacts.

Vehicle Extrication: A Practical Guide

Procedure 2
Relocation of a Front Seat with a Come-Along

Moving a front seat out of the way can be an important procedure and may be needed to access occupants on the other side of the vehicle. In most cases, the seat is not moved with an occupant in it because the sudden jarring movement could aggravate injuries. To move the seat back, the master link of a chain set is passed beneath the seat from the front to the rear.

Before working beneath the seat, the floor beneath the seat should be checked for obstructions and objects that could injure a crewmember such as a gun, knife, or needle.

If the area is clear of problems, the master link is pushed as far rearward as possible.

Interior Procedures

Procedure 2 *(continued)*
Relocation of a Front Seat with a Come-Along

The master link is grasped with the other hand and pulled the rest of the way to the rear seat floor area.

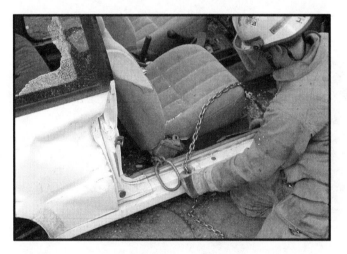

The master link is positioned at the outside rear corner of the seat. The chain is brought over the front corner of the seat and shortened at the master link with the chain shortener assembly. Using a sling hook and choking the seat may, at first, seem like a quicker way to rig the seat but actually can cause problems when tension is put on the system to pull the seat back. Instead of pulling the entire seat back, the choke chain will tighten around the seat bottom, wasting time and effort.

The block of the come-along is attached to the master link.

Vehicle Extrication: A Practical Guide

Procedure 2 *(continued)*
Relocation of a Front Seat with a Come-Along

The come-along is ultimately run through the interior of the vehicle, but cable can be let out easier and with more control from the outside. To pay out the right amount of cable, the tool operator walks the come-along back to the rear bumper area.

With the proper cable paid out, the come-along is passed through the interior of the vehicle and out the back window. When the come-along reaches the rear bumper area, it is attached to the anchor chain.

After the come-along is attached to the anchor chain, the officer directs the slack to be taken out of the line. Crewmembers check the rigging to make sure all the hooks and chains are connected in the proper manner.

Interior Procedures

Procedure 2 *(continued)*
Relocation of a Front Seat with a Come-Along

The anchor chain is supported by a ladder crib or other cribbing to prevent the chain from collapsing the trunk or liftback metal.

Initially, the seat bottom will collapse from the rearward force of the come-along. The come-along operator will find that relocating a seat with a 2-ton come-along is one of the most challenging come-along procedures performed on an automobile. In most cases, the come-along will break the seat loose just as it approaches its maximum pulling capacity. It is not uncommon for the come-along operator to need relief from a fresh crewmember to finish that last couple inches of pull.

When the seat breaks loose, occupants and crewmembers in the vehicle should be protected from springs that may release under tension as the seat is pulled rearward. To totally clear the area, the seat can be pulled by the come-along deep into the rear seat area.

Procedure 3
Interior Spread with a Winch

The procedure of using a come-along to expand and unfold an interior can be modified slightly and performed by a vehicle-mounted winch with similar results. Instead of attaching the chain to the "B" post, the chain on the pulling side of the vehicle is hooked to strong points of the chassis. The chains used should all be rated for use with the winch.

Both master links are connected to the winch cable hook.

As with the come-along, it's best to keep the cable snugly wound on the winch drum to avoid damaging the cable. To help control how the cable is wound on the drum, the cable is fed through a device called a fairlead. Without the benefit of a levelwinder, the cable may wind on the drum unevenly during its operation, and may have to be removed and rewound on the drum evenly after the emergency.

Interior Procedures

Procedure 3 *(continued)*
Interior Spread with a Winch

When operating a winch, the remote control allows the operator to move out of the hazard area in line with the cable. While chains will simply fall when broken, the spring effect of the cable can cause it to fly towards the winch if it were to break.

Careful monitoring of the victim, the obstruction/anchor and the chains is very important for a safe and successful operation. An 8000 lb winch is capable of lifting a vehicle off the ground, a situation that should be avoided. Careful and intelligent application of power, especially in this type of situation, can lead to good results while lack of situational awareness and disregard for safety can lead to serious problems for the victims and crewmembers alike.

As the interior is expanded, the inside crewmember should monitor the victim's disentanglement and stop the procedure if undesirable results start to occur. Even when desirable results are being achieved, constant monitoring to the vehicle, chains and anchor is important to avoid pulling the vehicle apart and causing failure of one of the components.

Vehicle Extrication: A Practical Guide

Procedure 4
Moving a Seat Rearward with Spreaders

Severe frontal collisions can leave occupants out of their seats and on the floor if unrestrained. Some crash configurations can result in the occupant being partially trapped on the floor and under the dashboard. When attempting to extricate the victim, you may have to forcefully push the front seat rearward.

A sturdy portion of the front hinge column is selected for a spreader tip to push against. The area of the hinge usually will be a strong area because of the reinforcement needed to support the weight of the door.

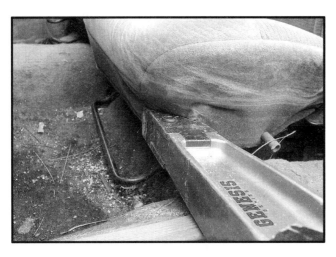

The other tip is carefully positioned against the front corner of the seat. To verify the location of the steel in the corner, the tool operator can press against the corner with a gloved hand to identify the perimeter of the seat structure.

Interior Procedures

Procedure 5
Removing a Front Seat with an Air Chisel

Removing a front seat with an air chisel can be accomplished but is difficult because of the visual obstructions encountered when trying to position the bit and then controlling it while it cuts. Generally the seat will be attached to the floor with four mounting brackets, one in each corner, or an adjustable rail. To gain access to the inside front and rear attachment points, the front outside bracket is cut away first.

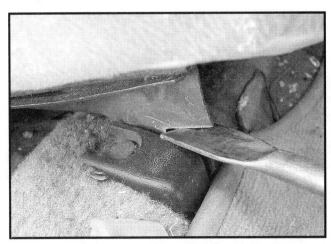

The curved chisel bit is the best choice for cutting though the bracket. The bit should be positioned so that it cuts from front to rear without entering the floor area or other area that will complicate the cut.

If a long chisel bit is available (18 in.), access to the outside rear bracket is possible even if the seat can not be moved upward to provide access.

Vehicle Extrication: A Practical Guide

Procedure 5 *(continued)*
Removing a Front Seat with an Air Chisel

Even with the long air chisel bit, complete cuts without close examination will be difficult. To help identify parts of the mounting bracket that are not cut, a Halligan can be used to pry up on the outer part of the seat.

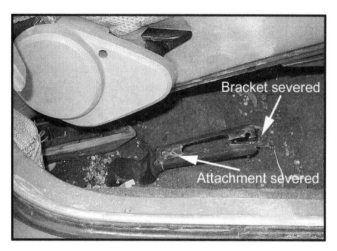

Mounting brackets are different in all vehicles, making it impossible to develop any specific method of seat removal. Fortunately, a high-performance air chisel is capable of making multiple quick cuts until the seat is released from the base bracket.

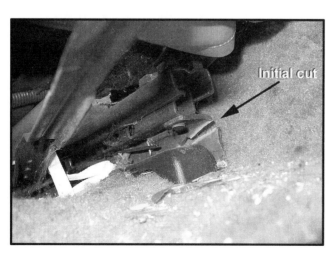

This interior front mounting bracket looks fairly complicated and confusing, but the cut can be simplified by avoiding all the sliding hardware and concentrating on the simple floor attachment point.

Interior Procedures

Procedure 5 *(continued)*
Removing a Front Seat with an Air Chisel

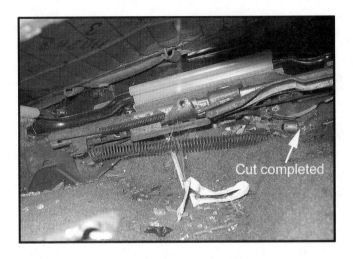

After the third cut is completed, the structure of the seat will be weakened, allowing it to be lifted for better access to the interior rear mounting bracket. Caution should be used as the springs become extended beyond their normal operating range. If enough stress is placed on them, they can become disconnected from the hardware and become airborne. The best way to handle them is to quickly cut through the springs to release them.

As with the front bracket, the sliding hardware of the rear seat bracket should be avoided and effort placed on severing the more solid and simpler floor mounting hardware.

Once the seat is removed, crewmembers will have good access to the opposite side front seat victim and rear seat victims.

Vehicle Extrication: A Practical Guide

Procedure 6
Removing a Seat Back with Hydraulic Cutters

To remove the seat back from a bucket seat, two approaches can be used. In both cases, the arm that supports the upper half of the bucket seat will be severed. The first approach is from the seat side with the cutters lined up between the seat back and seat bottom. The second method requires an approach from the front or rear of the seat. The blades are opened fully then positioned to wrap around the seat back assembly.

The shape and location of the seat back assembly will be the main factor in determining the best approach. The advantage to the forward/rear approach is that the seat assembly will be positioned deeper in the blades, creating a more powerful cut and less stress to the tool.

After the outer support arm is cut, a crewmember from the opposite side of the vehicle pulls on the top of the seat to expose the inner support arm. After the second assembly is cut, the seat back can be removed from the vehicle.

Interior Procedures

Procedure 7
Removing a Front Seat with Hydraulic Tools

Hydraulic spreaders and cutters can be used to quickly remove a front bucket seat or bench seat by first spreading the seat away from the floor then cutting the mounting hardware if necessary. The spreaders are first used to break any brackets and to provide unobstructed access to the remaining brackets. To start the spread, the tips are placed between the floor and the center outside edge of the steel seat frame.

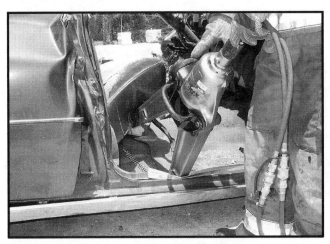

The seat frame is sturdy and holds up well to the spreading operation with little bending. As the spreaders are operated, springs attached to the underside of the seat will be extended beyond their normal operating length, posing a slight hazard to personnel in the area.

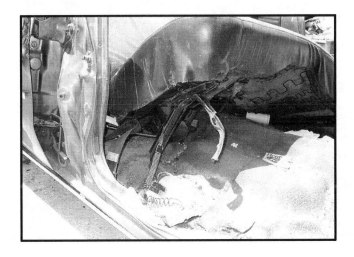

Both front and rear brackets may break during the spreading operation, but if only one breaks, the other can be cut with cutters. Seatbelts may need to be cut before the seat can be removed from the vehicle.

Procedure 8
Relocating a Steering Ring with Bolt Cutters

Sometimes just a few inches can make the difference between whether or not a victim can be removed from the driver's seat area. In these situations, a set of bolt cutters can be used to sever the steering ring and move it out of the way, or a second cut can be made and a portion of the ring can be completely removed. For maximum effectiveness, the ring should be cut as close to a spoke as possible.

After the cut is made, the ring can be pushed toward the center of the steering wheel. There is a natural tendency to pull the ring away from the center, but this should be avoided as it will only create a larger obstruction in most cases.

Once the ring is pushed toward and beyond the center of the wheel, a crewmember from the other side of the vehicle can pull on it to move it further away from the victim.

Interior Procedures

Procedure 9
Relocating a Steering Column with a Come-along

If it is necessary to relocate a steering column instead of using the preferred dash roll or dash lift procedures, the come-along offers plenty of power to complete the task. Rigging the column for maximum pulling leverage may seem overly complicated, but it is always better to have a little too much leverage than not quite enough. To start the chain wrap, the master link is inserted through the steering wheel from the front toward the rear (toward the victim.)

The master link is then passed through the top opening, and attached to the come-along block. The master link is hooked to the come-along early in the procedure simply to help hold the chain in place and to prevent it from falling into the driver's foot area. The sling hook is then passed over the top of the column, then under the column, back toward the rigger.

The sling hook is again passed over the top of the column and brought to the area beneath the column.

Procedure 9 *(continued)*
Relocating a Steering Column with a Come-along

As the hook reaches the area beneath the steering column, it is redirected through the opening in the bottom of the steering wheel. A mirror image of the first chain is starting to become apparent.

The hook is passed from rear to front through the top opening.

Once passed through the top opening, any excess chain is pulled through and will be removed by the chain shortener.

Interior Procedures

Procedure 9 *(continued)*
Relocating a Steering Column with a Come-along

The chain shortener is used to remove any slack chain and to create an equal pull on both sides of the chain. The excess chain is located so it won't interfere with the pulling operation when it begins. The master link is positioned on top of a slider crib to give it a better pulling angle and height and to decrease the likelihood of the chain getting tangled in the dashboard. The slider crib consists of two 4x4s positioned over the firewall area, with a third piece on top that will slide forward with the master link as the cable is taken up.

The leverage being created by the rigging can be seen in this photograph. The chain wraps on the lower part of the column help secure the chain in place, while the chain wrapped through the steering wheel creates leverage.

With the rigging completed, the come-along is walked to the anchor chain located at the front of the vehicle.

Vehicle Extrication: A Practical Guide

Procedure 9 *(continued)*
Relocating a Steering Column with a Come-along

The final positioning of the master link and slider crib is important to avoid binding. The 4x4s should be positioned so they are extending over the dashboard, fire wall, and rear hood area. The master link fits well when placed directly on top of the top 4x4. If, instead of the master link, the come-along block were to be placed on top of the 4x4, the cable would be pulling off the side of the pulley, creating an undesirable loading situation.

It is important to maximize the amount of cable used to perform the pull. This is a critical step especially when working on small vehicles with short front ends. To maximize the pulling potential, the come-along should be positioned over the front edge of the hood. As the slack is taken out of the system, the come-along invariably moves toward the driver, giving up valuable inches of pull. The key to all come-along operations is to have short chains and long cables.

When the rigging is completed, the officer directs the slack to be taken out of the system, then has the crewmembers check the hooks and connections to confirm that nothing has moved out of position as the slack was taken out. The come-along is then operated until the desired clearance is achieved. Considerable leverage on the steering column assembly is achieved in this procedure, causing concern by some that the steering column may break loose at the floor board swing upward injuring the driver. Because of this potential problem, this procedure should not be used as a first choice.

Interior Procedures

Procedure 10
Relocating a Steering Column with a Hydraulic Spreader

Steering column relocation procedures are not performed much anymore but occasionally may be required when other options such as the dash roll-up with rams or a dash lift with the spreaders are not possible. There are three distinct work areas in this operation, requiring three crewmembers. The first work area is the steering column and wheel. The crewmember in this area will rig the column with the chain and attach it to the spreader.

A crewmember will configure the spreaders for pulling and operate the tool.

Another crewmember will attach the anchor chains, set cribbing, and attach the chain to the spreader.

Vehicle Extrication: A Practical Guide

Procedure 10 *(continued)*
Relocating a Steering Column with a Hydraulic Spreader

The crewmember working with the steering column forms a loop around the steering column and attaches the hook to the chain. In the photograph, a grab hook is used, but sling hooks are just as common. The chain should be positioned as low on the steering column as possible for best results. If a sling hook is used, the open part of the hook should face the rear of the vehicle.

As the chain is pulled by the spreaders, the chain will tear through the lightweight dashboard with no movement of the column. This means that the spreaders are closing and using critical inches of closing distance with no impact. To minimize the wasted inches, a piece of cribbing is inserted between the chain and the speedometer. A 4x4 is less likely to break than a 2x4 when under load and should be used if it fits.

The chain is then brought to the spreader arm and connected. To create a better pulling angle, a slider crib is built under the chain. A slider crib is made from three 4x4s; two are placed like rails over the firewall area and the third is placed on top. The top piece is positioned at the rear edge of the bottom pieces. In addition to providing a good pulling angle, the slider keeps the chain out of the dashboard as much as possible. It gets its name by the way it slides forward on the bottom pieces of cribbing as the spreaders close.

Interior Procedures

Procedure 10 *(continued)*
Relocating a Steering Column with a Hydraulic Spreader

The crewmember assigned to the spreaders is responsible for attaching the pulling hooks to the spreaders. When attaching the hooks and chains, the manufacturer's procedures should be followed of course, but there are some procedures common to most tools. When attaching the hooks, they should be attached to form a mirror image of one another. It is easier to connect the chains to the hooks if the open part of the hook is facing up, making this a good standard procedure.

The way that the hooks are connected to the spreader arms varies, but if pins are used, it is believed best to insert the pins from the bottom. When inserted from the top down, the pins can be pushed out by the hook when the tool is placed on the hood of the vehicle.

The amount and location of damage to the vehicle will determine where the spreader operator stands; but if given a choice, the operator should stand on the passenger side of the vehicle. This position allows the spreaders to be laid on the hood as they are being hooked up and operated. If positioned on the driver's side, the tool operator would have to constantly hold the spreaders to prevent them from falling off the hood. Once in position, the spreaders are opened to their maximum opening distance.

Vehicle Extrication: A Practical Guide

Procedure 10 *(continued)*
Relocating a Steering Column with a Hydraulic Spreader

The crewmember responsible for the anchor chain selects a strong point on the chassis to attach the chain. The chains supplied with spreaders usually are not very long, so the chain should be attached toward the front end of the vehicle if possible.

This chain set is equipped with grab hook at the end; other manufacturers use sling hooks. The chain in the photograph is wrapped around a stout part of the chassis then hooked back into the chain. The loop formed by the chain and hook should be as short as possible to allow more chain to be available for hooking up to the spreaders.

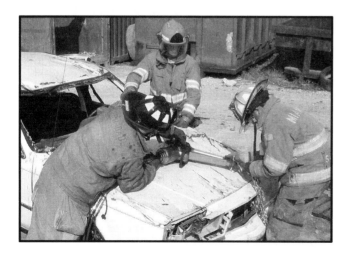

With the steering column chain hooked up, the crewmember assigned to the anchor chain pulls the spreader arm toward the front of the vehicle to align the spreader and chain. After aligning them, move the spreader in a straight line toward the anchor chain. Good communications and control are very important in this part of the operation. The tool operator must not operate the tool or have a hand on the control while the crewmember is hooking up the anchor chain. Inadvertent operation of the spreader could cause serious injuries.

Interior Procedures

Procedure 10 *(continued)*
Relocating a Steering Column with a Hydraulic Spreader

When attaching the anchor chain to the spreader arm, first remove the majority of the slack chain. An excessive amount of time or effort should not be used to remove one additional link of slack chain when making the hookup; that small amount of slack can be removed with cribbing.

After both chains are attached, attention can be given to removing slack from the chain and supporting the anchor chain. A ladder crib or other types of cribbing can be slid under the anchor chain to minimize any crushing of the front end. By reducing crush, the closing distance of the spreaders can be used to pull the column up instead of crushing the hood and grill down.

To remove slack from the chain, cribbing is applied in front of the grill area. Placing the cribbing in front of the grill removes more slack than if the cribbing were placed on top of the hood. Again, to avoid unintentional operation of the tool, it is very important that the tool operator not have a hand on the spreader control.

Vehicle Extrication: A Practical Guide

Procedure 10 *(continued)*
Relocating a Steering Column with a Hydraulic Spreader

After the slack has been removed from the system, the officer directs the tool operator to remove the last bit of slack by closing the spreader slightly. The crewmembers are directed to perform a final check of their parts of the system. With the final check completed, the officer directs the tool operator to make the pull. The spreaders are closed while monitoring the chains and the reaction of the vehicle components.

The importance of cribbing can be seen after the spreaders have closed. While some crush occurred, the cribbing prevented greater crushing to the front end.

The steering column is pulled up and away from the passenger compartment. The column should only be pulled as far as needed to remove the victim. Pulling the column more than needed will increase the risk of unwanted reaction of the vehicle components.

Interior Procedures

Procedure 11
Removing a Steering Wheel with Hydraulic Cutters

To increase interior space and work area, the steering column can be severed with hydraulic cutters. Some steering columns are rather stout, requiring very strong cutters to make the cut. To maximize the space created by the cut, the cutters should be positioned as low on the column as possible.

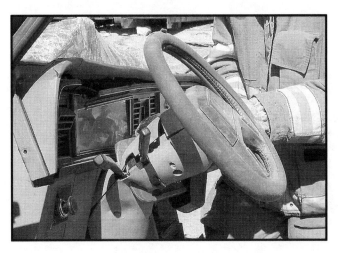

As the cutters close, the plastic trim around the column will break apart. If the cutters don't have the power to complete the cut on the first attempt, the plastic trim should be removed and the cutter repositioned with the steering column deeper in the jaws of the cutters. For maximum cutting force, the cutters should be positioned with the column in the notch, located deep in the cutter blades.

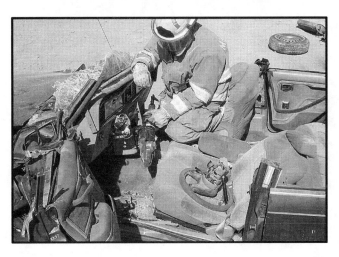

This procedure shouldn't be used when a victim is trapped behind or directly in contact with the steering ring, but it can be used when removal of the steering column is necessary to gain access to other parts of the vehicle. When cut, the steering column will snap with enough force to propel the severed assembly away from the remaining column. To prevent the wheel from hitting the occupant, a crewmember should grasp the wheel as it is cut. This photograph shows the location of the severed steering wheel and column immediately after it was cut.

Vehicle Extrication: A Practical Guide

Procedure 12
Relocating a Pedal with a Length of Rope

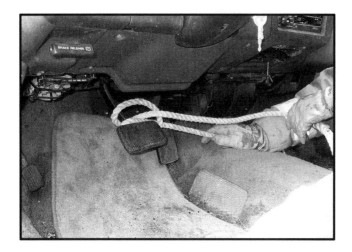

Pedals aren't designed to be strong when loaded from the side or when pulled upward, a fact that crewmembers can take advantage of when it's necessary to move a lightweight pedal. A utility rope and two crewmembers may be all that's needed to relocate the brake pedal of a lightweight vehicle. The set up is simple; the rope is run around the pedal arm and both ends of the rope are grasped by crewmembers outside the vehicle.

In a coordinated effort, the crewmembers pull the pedal in the desired direction. If the door adjacent to the crewmembers will open easily on its hinges, it can be used for additional gripping strength and leverage. With the door nearly closed, the rope is wrapped around the window post or the door. The door is then pushed/pulled open by the crewmembers.

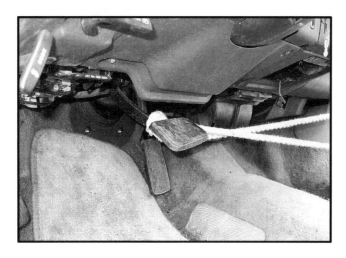

While mid-sized and full-sized vehicle pedals may be too large and strong to pull with a rope, it may be all that's needed on a compact vehicle.

Interior Procedures

Procedure 13
Relocating a Pedal Across the Hood with a Come-along

When it is necessary to pull a pedal straight up to free a victim's foot, the across-the-hood method is a good choice. The procedure is started by placing the come-along at the rear edge of the hood and then hooking the master link to the block. By hooking the master link to the block, the come-along will prevent the chain from falling into the vehicle as the chain is rigged.

Access to the pedal and positioning of the victim will determine how the chain is attached to the pedal. One method is to pass the sling hook around the pedal arm, and then bring the chain back to the chain shortener. This method provides a sure method of attachment and keeps the attachment points of the chain in clear sight.

Another method is to simply attach the hook to the pedal arm. This method works best if the sling hook has a latch to help hold the hook in place during the rigging process.

Vehicle Extrication: A Practical Guide

Procedure 13 *(continued)*
Relocating a Pedal Across the Hood with a Come-along

Choking is not a good option because the pedal arm is usually smaller than the length of the hook making it impossible not to side load the hook. Additionally, the chain won't be positioned against the thickest part of the hook where it is strongest.

To keep the chain out of the dashboard as much as possible, a slider crib is built out of three 4x4s. The come-along block should be positioned forward of the top 4x4 to avoid any possibility of side loading. The master link is positioned directly on top of the 4x4.

The come-along is attached to the anchor chain at the front end of the vehicle. After taking up the slack, the officer has the crew check the rigging before the pulling operation begins. The pedal should be watched carefully by a crewmember other than the come-along operator who won't be in a position to see the pedal moving.

Interior Procedures

Procedure 14
Relocating a Pedal through the Passenger's Door with a Come-along

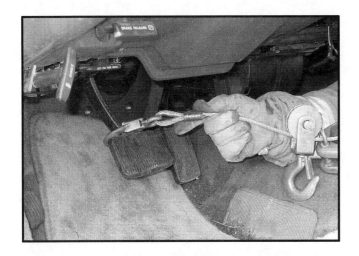

Usually a chain is used to connect a come-along to an object to prevent damage from occurring to the come-along cable. The procedure of relocating a pedal through the passenger's front door is one of the few times that it is not necessary to use a chain. If the size and shape of the pedal arm permits, the hook at the end of the cable can be connected directly to the pedal arm.

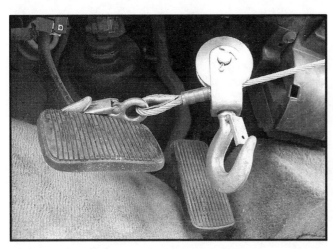

The come-along block, normally used when a 2:1 mechanical advantage is needed, simply rides on the cable unused.

The come-along is then walked out to the passenger side "B" post.

309

Procedure 14 *(continued)*
Relocating a Pedal through the Passenger's Door with a Come-along

The anchor chain is attached to the rearmost roof post after removing the glass. The chain is then shortened to position the anchor chain master link at the "B" post.

The come-along is attached and the slack is taken up. Before operating the come-along under load, check the chain connections to confirm that they are still in the proper position and connected.

Operating the come-along handle is awkward, and there may be benefit in reversing the handle in the come-along bale. Reversing the handle will result in less leverage due to the shorter usable length but may be a little easier to operate in the confined area. The importance of proper rigging of the anchor chain is evident. If the chain were rigged longer than shown in the photograph, the come-along would be further in the passenger compartment, making it difficult to operate.

Interior Procedures

Procedure 15
Relocating a Pedal through the Rear Windshield with a Come-along

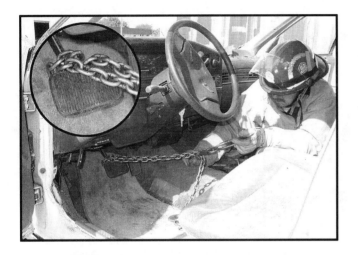

If a pedal needs to be pulled upwards to free a victim's foot, and the across-the-hood method can not be used, the come-along can be rigged to pull out through the rear windshield. The configuration of the pull will result in the chain being pulled toward the end of the pedal, so the chain should be securely wrapped to prevent it from slipping off the pedal.

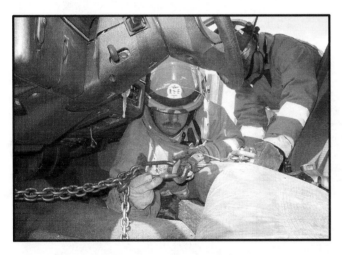

The come-along can be used in the 2:1 or 1:1 mechanical advantage configuration for this pull. In this case, the block was used to connect to the master link, giving it a 2:1 mechanical advantage.

For ease in letting the cable out to the correct length, the come-along is walked back a distance equivalent to the distance needed to reach the rear edge of the trunk lid. The interior of the vehicle is too congested to try to push and pull the come-along to the rear of the vehicle.

Vehicle Extrication: A Practical Guide

Procedure 15 *(continued)*
Relocating a Pedal through the Rear Windshield with a Come-along

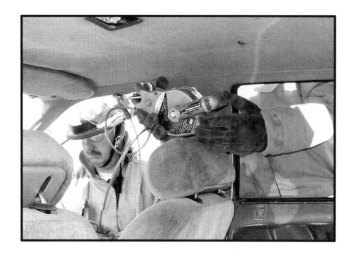

The come-along is then passed through the interior compartment of the vehicle until it is out the rear windshield.

The anchor chain is rigged so the master link is positioned at the upper rear edge of the trunk. The come-along is then attached to the master link. Relocating a pedal does not require much force, but ladder cribbing was placed on the trunk lid anyway to distribute the force created by the chains.

The slack is removed from the system, and all connecting points are checked to confirm that all hooks are properly loaded and connected. After checking the rigging, use the come-along to relocate the pedal. The come-along operator probably won't be able to see how the pull is progressing, so another crewmember will be needed to check the progress.

Interior Procedures

Procedure 16
Relocating a Pedal over the Roof with a Come-along

If it is necessary to pull the pedal upward and toward the left side of the vehicle, a chain hook is connected to the pedal, and the master link is positioned at the upper rear corner of the driver's window.

The come-along is attached to the master link and positioned to be walked back to the opposite rear corner of the roof.

The anchor chain is connected to the "C" post or other lower attachment point that will allow the master link to be positioned at the rear corner. The cable on the come-along is then paid out, and the master link of the anchor chain is connected.

Vehicle Extrication: A Practical Guide

Procedure 16 *(continued)*
Relocating a Pedal over the Roof with a Come-along

The anchor chain is kept as short as possible to maximize the amount of cable available for pulling. This anchor chain formed a small loop around the "D" post.

To minimize the amount of downward movement of the roof sheet metal, ladder cribbing or single pieces or cribbing can be placed on the roof.

The positioning of the anchor chain master link is important in this evolution for two reasons. First, if the anchor chain is rigged so the master link is positioned closer to the center of the roof than the edge, the come-along operator will have difficulty in reaching the handle. If, on the other hand, a low, chassis anchor point was used and the master link positioned over the side window, the come-along cables would be bent over the edge of the roof. The positioning shown results in an easy pull.

314

Interior Procedures

Procedure 17
Removing a Pedal with a Whizzer

An extrication scene is probably one of the worst places to use a tool that produces a lot of sparks because of the risk to the victim who would not be able to leave the area if a fire were to ignite. Nonetheless, it is important to have knowledge of all options at a crash scene, including the hazardous ones. If a whizzer is selected for use at a crash scene, proper fire suppression equipment should be in place before the procedure is started.

Positioning the whizzer to cut completely through the pedal arm may be difficult or impossible because of limited access to the pedal arm. If the pedal can be cut most of the way, an adjustable wrench may provide enough leverage to bend the arm out of the way.

If enough leverage is gained with the wrench, the metal will become fatigued and may break off completely. While the pedal can be removed with the whizzer, it may require a couple of air cylinders to complete the time-consuming task. The time factor and spark hazard make it a possible but not an optimal procedure for removing a pedal.

Procedure 18
Removing a Pedal with a Manually-operated Hydraulic Cutter

Manually operated pedal cutters can be used to cut pedals when access to the pedal area is limited. Their small size and light weight allows them to be positioned with little effort.

The manually operated pumps are simple devices that should be operated on level ground. Some pumps are equipped with an attached base to help keep the pump upright and flat.

Most manually operated pumps have a valve that is moved to the "closed" position to operate whatever tool is attached to the pump. The tool is then released by opening the valve. Some manual hydraulic pumps also have vents that need to be opened when operated.

Interior Procedures

Procedure 18 *(continued)*
Removing a Pedal with a Manually-operated Hydraulic Cutter

To cut the pedal arm, the cutter is positioned with the entire arm in jaws of the cutter.

While holding the cutter in place, the crewmember operates the pump handle, which in turn causes the cutter blades to close.

The crewmember continues to operate the pump until the pedal is cut away. As the cut is completed, the pedal will usually fall gently to the floor. If possible and if desired, the pedal can be held as it is cut to prevent uncontrolled movement.

Vehicle Extrication: A Practical Guide

Procedure 19
Removing a Pedal with Hydraulic Cutters

Full-size hydraulic cutters can be used to cut pedal arms, but visibility of both the pedal and the occupant's foot may be obstructed by the tool itself. For maximum cutting power, the pedal arm should be placed in the cutter notch located deep in the cutter jaws.

Specially designed pedal cutters are more maneuverable than full-size cutters because of their light weight, making them and manually hydraulic cutters better tool choices when pedals need to be cut.

Their small size allows the cutters to be placed high on the arm, creating more space for further disentanglement and removal.

Interior Procedures

Procedure 20
Using a Hi-Lift Jack to Create Interior Space

Some crash configurations that crews encounter may require the interior of the vehicle to be pushed outward to free or disentangle the victim. If attempts to pop a door open causes the door to move in on the victim and other procedures are indicated, the Hi-Lift jack may be used to spread the interior apart. If the jack is used in close proximity to the victim, the spreading operation will require extremely careful monitoring to avoid any sudden movements or releases of force. These photographs illustrate an extreme case, requiring careful operation of the jack very close to the victim.

The jack's operating lever is placed in the "raise" position before putting it in position.

The movable end of the jack is positioned in a strong area. Cribbing can be used to further distribute the load, minimizing the chance of the metal bending.

Vehicle Extrication: A Practical Guide

Procedure 20 *(continued)*
Using a Hi-Lift Jack to Create Interior Space

When spreading an interior in a small automobile, the full length of the jack may be used, requiring the use of cribbing to increase the amount of usable spread.

The base of the jack is placed against the area that needs to be moved away from the victim. As the jack is positioned, it should be kept away from the victim as much as possible in case it slips during the spreading action. Like other procedures that place a tool and the victim in close proximity, spreading with the Hi-Lift jack requires careful consideration and forecasting to accurately perform a risk/benefit analysis before the procedure is carried out.

In this situation, the door was forced away from the victim, unlatching in the process. Another option would have been to operate the Hi-Lift until the door was moved into a position that allowed another tool to be used that would not operate as closely to the victim.

Interior Procedures

Procedure 21
Interior Spread with a Hydraulic Ram

Hydraulic rams, especially the three piece telescoping rams, provide a solution for the problem of side to side interior crushing that occurs during severe side impacts. In this example, for clarity, the roof has been left on the vehicle, while in most crashes of this severity crewmembers would probably choose to remove the roof for better access to the victims and visualization of the entrapment

If the objective is to spread the "B" posts and pillars apart, the tool operator may need help positioning the ram as there won't be very many positions to choose from.

In this example, the base of the ram is positioned at the edge of the roof and the top of the "B" post.

Procedure 21 *(continued)*
Interior Spread with a Hydraulic Ram

The head of the ram is positioned against a sturdy part of the "B" pillar. Once in position and extended slightly to snug it up, the tool operator should check the base and head to confirm their proper position.

In this example, as the ram extended, the base pushed outward toward the driver's side because of less resistance on that side of the vehicle. Throughout the procedure, the tool operator should be alert to any chance of a door popping open as the ram is extended. A door popping open suddenly could result in injury to the tool operator if positioned in front of the door.

Once the driver's side extended outward fully, resistance was created, causing the head side of the ram to push the passenger side outward. The advantage of the three-piece telescoping ram can be seen by comparing the last three photographs in the series.

Interior Procedures

Procedure 22
Special Interior Problems in Police Cars

Full-size police cars used to transport prisoners can pose additional problems for crews responsible for the extrication of occupants. While not apparent from the exterior, the dividers that separate the front and rear seats are large and have heavy framework.

The framework of the divider is made of steel tubing that supports a lower metal panel and an upper panel that can be made of Plexiglas, Lexan, or a metal mesh. The system is designed to protect the police officer from both physical and biohazard assaults. The divider is attached to the floor and center pillar or post with brackets that are bolted or screwed into place.

The top edge of the divider may have a small gap and probably won't have any attachment points to the ceiling.

Procedure 22 *(continued)*
Special Interior Problems in Police Cars

The upper part of the divider is supported by a steel bracket that is attached to the "B" post. If the support bracket is cut in two spots, the top of the cage will be released and the remaining bracket won't obstruct the work area.

Older versions of dividers utilize the shoulder harness bracket in the "B" post as an attachment point for the bracket, but new versions use brackets that are clamped to the interior of the "B" post.

The lower portion of the divider may be screwed or bolted to the floor or may use a bracket that attaches to the front seat track. Dividers that further divide the rear seat into left and right sides may use a lower bracket mounted to the transmission hump. The tubing at the base of the cage can be cut or an attachment plate can be cut with an air chisel.

Interior Procedures

Procedure 23
Performing a Dash Roll with a Hi-Lift Jack

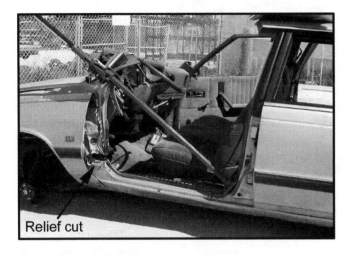

A Hi-Lift jack can be used to perform a dash roll in the same way a hydraulic ram is used. To weaken the structural area that joins the rocker panel to the front hinge pillar, a relief cut is made with an available tool. The relief cut will typically extend about 6–8 in. into the forward hinge pillar and can be made with a hydraulic cutter, an air chisel, or reciprocating saw.

The base plate of the jack is positioned at the bottom of the center pillar. The base plate of the jack is rather large and will distribute the load over a large area, reducing the chance of the metal tearing as the jack is operated.

The nose of the jack is positioned in the area where the "A" post becomes the front hinge pillar. This curved area is stronger than the bottom of the "A" post itself and less likely to bend when the jack is operated. The jack is operated until the victim is freed or until the entire usable length of the jack is utilized. If the jack must be removed to access and remove the victim, cribbing should be built beneath the vehicle to support any sections of the vehicle that moved upward when the jack was operated. As the jack is retracted, the vehicle may move downward if not cribbed, causing the vehicle to close in on the victim again.

Vehicle Extrication: A Practical Guide

Procedure 24
Dash Roll-up with Hydraulic Tools

During the dash roll-up, the dashboard will be pushed upward and forward away from the occupant compartment. For the dash to roll, the roof must be disconnected from the forward hinge pillar, or the roof and posts would hold the dashboard back, preventing the roll. A roof flap or removal is often the most productive way to sever this connection, but simple severing, or removal of the "A" post provides the same result but without the benefits derived from the other procedures. To sever the post, one cut is all that is required.

While one cut may sever the connection between the roof and the forward hinge pillar, there is a chance that the dashboard will bind against the post as the dashboard moves up and forward. To eliminate this problem, a second cut can be made.

With the post removed, the dashboard can move freely without binding on the post.

Interior Procedures

Procedure 24 *(continued)*
Dash Roll-up with Hydraulic Tools

Like the roof, the rocker panel needs to be separated from the forward hinge pillar if the dash is going to roll forward. To weaken the structure and cause the separation, a relief cut is made below the bottom hinge, parallel to the ground. The relief cut should not be angled downward into the rocker panel.

The next step is optional but increases the chance of success when performing the operation. To weaken the structure further, the top rail is severed, allowing the dashboard and firewall to move independently from the rest of the front end of the vehicle. The first step in cutting the top rail is to gain access to it by removing the fender. This is done by placing the tips of the spreader between the hinge pillar and the fender and spreading.

The fender will break away from its bottom attachment points quickly and flip upward easily. The spreaders are repositioned as necessary to remove most of the fender.

Vehicle Extrication: A Practical Guide

Procedure 24 *(continued)*
Dash Roll-up with Hydraulic Tools

With the fender out of the way, the top of the top rail is exposed by spreading between the top rail and the hood.

After the hood has been spread out of the way, the top rail is visible. If the top rail isn't severed, the entire front end of the vehicle remains rigid and in one piece, causing resistance to the rolling movement of the dashboard and firewall.

If the vehicle is equipped with struts in the front suspension, the top rail is cut just rearward of the strut assembly. If the vehicle doesn't use struts, the cut is made in the same basic area, a foot or two forward of the firewall. The top rail should be completely severed.

Interior Procedures

Procedure 24 *(continued)*
Dash Roll-up with Hydraulic Tools

After the cuts are made, the ram is placed in position. Some rams have dual pistons that extend from both ends of the ram while the other more common type of ram has one. If a one piston ram is used, the larger part of the ram (base) is placed low in the lower rear corner of the door, and the movable piston part of the ram is placed in the upper position. This is done for easy handling and good balance.

Final positioning of the ram head is based on the choices available after sizing up the damage to the hinge pillar. Generally, the head is placed in the curved area where the "A" post connects to the hinge pillar. If the door is left on the vehicle, the top hinge often provides good support for the ram.

After the process of weakening some structures is completed, actions are taken to strengthen others. The base of the ram will be placed at the lower rear corner of the door opening, an area constructed of only sheet metal. The area is not exceptionally strong and often buckles when force is exerted on it by the end of the ram. To give the area some support, cribbing is placed beneath the corner.

Procedure 24 (continued)
Dash Roll-up with Hydraulic Tools

Even with cribbing placed beneath the ram base, the metal may buckle and fail because of damage, rust, or lightweight construction. In the worst cases, the ram head will stay stationary, and the base will push through the rear pillar, resulting in no movement of the dash.

The best method for providing a solid push-off point for the ram is to use the spreader tips. Before using this procedure, the tool manufacturer should be contacted to determine if the procedure is recommended. To use the spreader as a push-off point, the ram is put in the desired position, and the spreaders are clamped down on the rocker panel as tightly as possible. If they are not clamped down tightly, the ram will cause the spreader arms to bend out of alignment, possibly causing them to break.

To enlarge the surface area of the ram base, accessory plates can be used. By doubling or tripling the size of the contact area between the ram base and the metal, the metal is less likely to tear.

Interior Procedures

Procedure 24 *(continued)*
Dash Roll-up with Hydraulic Tools

If the required equipment is available, two rams can be used at the same time. Good coordination is needed because movement of one ram will have an impact on the other.

The ram is extended to the distance needed to remove the victim plus a little more to compensate for any settling that will occur if the ram is retracted and removed to access and remove the victim. As the ram extends, the area beneath the relief cut will have a tendency to move upward. This area is cribbed to prevent settling that will occur when the ram is retracted.

A two-sided dash roll was completed on this vehicle; one side had the door removed while the door was left intact on the other side. Either procedure is acceptable. Leaving the door attached creates an obstruction but the hinge pillar remains intact because no hinge work is required to remove the door.

Procedure 25
Dash Lift

The dash lift, like the dash roll-up, is a good procedure to use when the dashboard and firewall assembly have been pushed rearward onto the occupants. The difference between the dash roll-up and the dash lift is the direction the dashboard is pushed to free the victims. The dash lift pushes the dashboard upward, instead of forward as the dash roll does, making it ideal for situations when the front of the vehicle is against an object that would resist forward movement.

Like the dash roll, steps are taken to weaken some structures and strengthen others. While the steps may seem to take up critical time, omission of the steps may result in the victim remaining trapped. Additionally, if steps to weaken the structure are skipped, parts of the vehicle may tear or be moved out of position complicating the situation. Once a piece of metal has been bent, it is usually impossible to undo the act. One of the important step starts with flattening the fender just to the rear of the strut assembly.

As the spreaders flatten the fender, it will pull away from the door, exposing the hinges. The flattening of the fender will also make it easier to cut the fender as shown in the following steps.

Interior Procedures

Procedure 25 *(continued)*
Dash Lift

Initially the bottom hinge is exposed, but the top hinge will also become exposed after the fender is removed.

The next cut will accomplish two things. First, the fender is severed into two pieces, making it easier to remove the rear portion. Additionally, the first of a series of cuts is made to sever the top rail assembly. As the first cut in the fender is completed, the cutters are placed as deep as possible in the first cut and a second cut is made.

The relationship between the strut and the cut can be seen inside the fender. If the vehicle does not use a strut assembly, the cut is made a foot or so in front of the front hinge pillar.

Procedure 25 *(continued)*
Dash Lift

To remove the rear portion of the fender, the spreaders are inserted in the gap between the door and the fender, and the spreaders are opened.

The fender may tear away completely, or it may fold upward. With either result, the fender will be out of the way.

To further weaken the structure that would resist the dash lift, additional cuts are made in the top rail to completely sever it.

Interior Procedures

Procedure 25 (*continued*)
Dash Lift

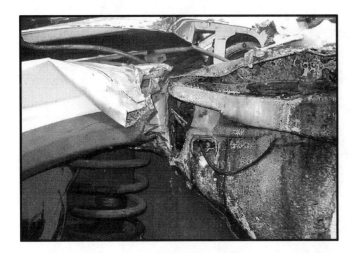

The top rail has been completely severed in this photograph.

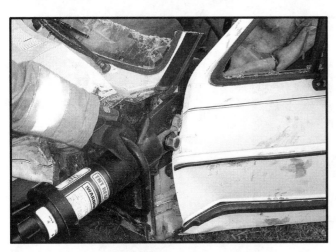

The door is removed next, preferably with a hydraulic cutter so damage to the front pillar is minimized. If the door weren't removed, it would act like a splint holding the front hinge pillar together and complicating access to the pillar.

The next cuts will accomplish two things. First, the forward hinge pillar will be severed, allowing the dashboard to move upward and away from the bottom half of the hinge pillar. Additionally, a gap will be created for the placement of the spreader tips.

Procedure 25 *(continued)*
Dash Lift

The first cut is made about halfway up the hinge pillar, as deep as possible.

The next cut is made about 3 in. above or below the first cut.

After the second cut is made, the cutters can be rotated deeper into the hinge pillar to make a third cut and create additional severing and weakening of the structure. This cut may be made between the first and second cuts, resulting in removal of the metal between the two cuts. If this metal does not fall away after the third cut, it will need to be bent out of the way by grabbing it between the spreader tips and then bending it forward to open up the gap between the first two cuts.

Interior Procedures

Procedure 25 *(continued)*
Dash Lift

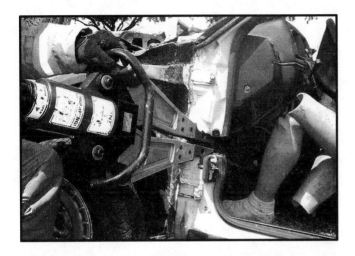

After the metal between the cuts is removed, cribbing is placed below the forward hinge pillar. The cribbing will provide strength and support in the bottom portion of the pillar. The tips of the spreader are placed in the gap that was created with the cutters.

To avoid buckling of the pillar, the tips should be positioned forward of the rear edge of the pillar.

The spreaders are operated and the dashboard moves upward and away from the victim. Again, while similar to the dash roll-up in some respects, the direction of movement of the dash is upward instead of forward as is the case with the dash roll-up.

·7·

Crash-Related Impalement

Crews may occasionally have to treat vehicle occupants who have received injuries from road debris or structures that were struck as their vehicle left the road.

Fig. 7–1 A near-miss illustrating how easily an impalement problem could occur.

If an object that is long and slender strikes an occupant, the object may impale the occupant. Crew members working at this type of incident will encounter several problems that need to be addressed before the victim is transported to the trauma center or hospital.

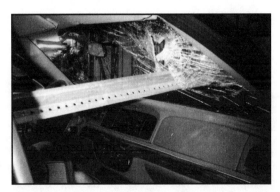

Fig. 7–2 Another view of how easily an impalement problem could occur.

As with every crash, a good size-up is needed to identify the problems before any solutions can be developed. When sizing up an impalement, the crew should realize that the procedures they perform on the scene will have an impact on the victim, the unit providing transportation to the hospital or trauma center, the emergency department staff, and the surgeons in the operating room.

Fig. 7–3 Another example of how easily an impalement problem could occur.

As the size-up begins, one of the primary concerns is the victim's medical status. The crew responsible for the medical care of the patient should evaluate and begin treatment of the victim while the extrication crew develops a plan for removing the victim from the vehicle. If the victim has expired, local protocol should address how the scene is handled and how the crew should interact with law enforcement agencies on the scene.

If the victim survives the impalement, the medical crew should examine the victim to determine the extent of the victim's injuries and the degree of impalement. During the examination of the victim, the medical crews will determine if the object has entered one part of the body and exited another or if the object entered only.

While the medical crew is conducting an examination of the victim, the extrication crew should determine if the victim can be transported to the local medical facility with the object intact or if the object will have to be reduced in length and weight. If the object needs to be reduced in length, a decision will have to be made about how much of the object should remain and how much should be cut away. The first factor in making that determination is the space needed to remove the victim from the vehicle. As the victim is moved out of the vehicle, the object will have to clear the roof, door, and any other part of the vehicle that the victim will pass on the way to the stretcher.

The next consideration is how the victim will fit in the ground transport unit or helicopter. If the stretcher in the ground unit locks against the wall while the vehicle is in motion, the distance between the victim and the wall should be considered. The same holds true for transportation in a helicopter. The interior size of helicopters can vary quite a bit, and if unsure about the available space, the air crew should be consulted while they're flying to the scene. The staff in the operating room will be the last to handle the object before it is removed, and enough of the object should remain so the surgeons will be able to get a good grip on the object during surgery.

If the crew determines that the object will have to be cut down in length, the proper tool will have to be selected. One of the first tools to come to mind will probably be the hydraulic cutter. The hydraulic cutter is easy to use and powerful, but there is one drawback to the cutter that should be considered before the plan is firmed up. If the cutter is used on anything but a solid object, like a piece of reinforcing steel (rebar), the shape of the object is likely to change as it is cut. A dramatic change in shape may or may not be a problem for the surgeons, but if given a choice, it would be best not to change the shape unnecessarily. Local protocol and policy may address this subject and should be considered the best source of information and direction.

Crash-Related Impalement

Aluminum sign poles

Steel sign posts

Galvanized pipe used to keep pedestrians and bikes away from the guard rail.

If struck from the end, the divider could make contact with the vehicle.

Trees and wood debris on the road.

Fence poles are commonly encountered by vehicles as they leave the road.

Fig. 7–4 Potential roadway implement hazards

Other tools that may be useful in shortening an impaled object include reciprocating saws, pneumatic whizzers, and rebar cutters. If given a choice of tools, the impact of vibration and heat generation should be given consideration. Reciprocating saws can operate smoothly, but if the blade binds a jarring motion will occur. The whizzer may be able to cut objects that other tools can't cut, but it produces an endless shower of sparks as it cuts. Both tools produce heat, the whizzer more than the saw because it grinds metal away with its abrasive wheel instead of cutting it away with a blade. As the object is cut and heated, some of the heat will be transferred through the metal to the victim. To decrease the amount of heat transfer, the location of the cut can be moved further away from the victim, and/or wet sterile towels can be placed between the victim and the cut to cool the metal. Heat-Fence, a commercially available product used by welders, can be packed around the object between the victim and the cut to reduce heat.

After the object is cut, the Heat-Fence can be removed and any residual wiped or washed away at the hospital. The emergency department should be notified that the product was used and of the need to rinse any remaining material away.

Before cutting the object, crew members should be positioned to grip and support the object as needed, to prevent it from making any sudden movements when cut.

Fig. 7–6 A crewmember should support and steady the object as it is cut to prevent movement both during and after the cut is made.

Once the object is cut down in size, it should be packed and supported to minimize any movement during transportation to the hospital or trauma center. Throughout the entire process crew members responsible for the extrication and medical care should take steps to minimize any movement of the object that could cause additional injury or internal bleeding.

Fig. 7–5 A good size-up will prevent surprises when preparing to move a victim to a stretcher. As with other penetrating injuries, the victim should be examined for both entry and exit wounds.

PROCEDURES

Vehicle Extrication: A Practical Guide

Procedure 1
Cutting Rebar with Hydraulic Cutters

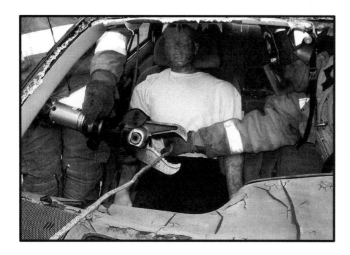

If the middle portion of the cutter is used to sever an object, the object is likely to move as it is cut.

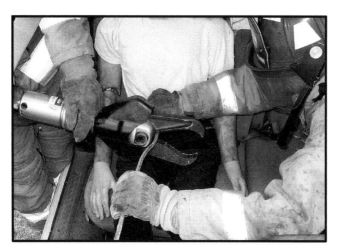

To prevent movement, the object is placed in the notch at the root of the cutter, and the object is held on both sides of the cut. Despite the appearance in the photograph, the crewmember's right hand is several inches from the cutter blades.

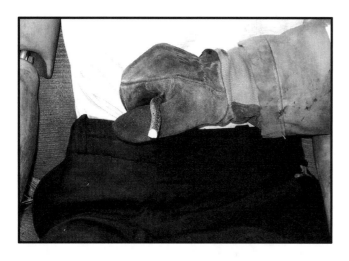

The notch completed the cut with little movement or change in the shape of the rebar.

Procedure 2
Heat-Fence—A Welder's Product that Reduces Heat Transfer

Heat-Fence comes in a tub and can be easily stored with a torch set or other tools.

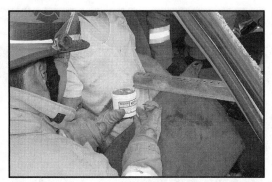

Heat-Fence has the consistency of wood putty and can be removed from the tub by hand.

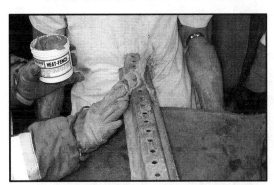

The Heat-Fence is placed between the victim and the anticipated cutting location.

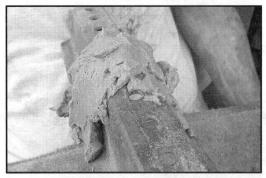

A generous amount of Heat-Fence is applied to minimize the heat transfer to the victim.

The cut is made beyond the edge of where the Heat-Fence was applied.

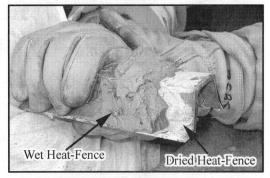

As it absorbs heat, the Heat-Fence dries and turns from gray to white.

Procedure 3
Cutting Rebar with a Reciprocating Saw

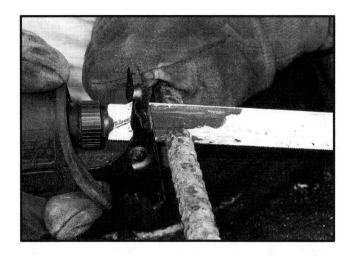

To reduce bouncing when cutting, the foot of the reciprocating saw should be placed lightly against the object as it is cut.

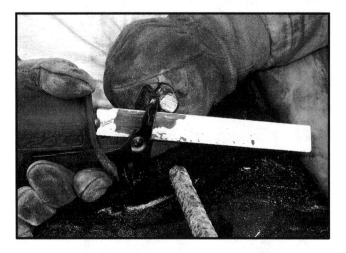

Like the hydraulic cutter, little deformity occurs with the saw, but some heat is generated.

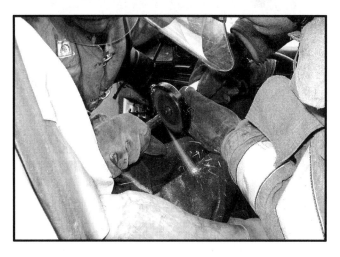

The whizzer produces a lot of sparks and metal fragments when operated. The fire danger it presents requires it to be used very carefully and with adequate preparations in the event a fire does start.

Crash-Related Impalement

Procedure 4
Cutting Large Pipes with a Reciprocating Saw

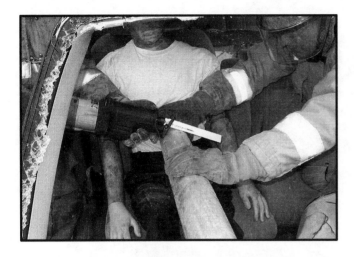

For large aluminum pipes, the reciprocating saw with its selection of long blades is a good choice.

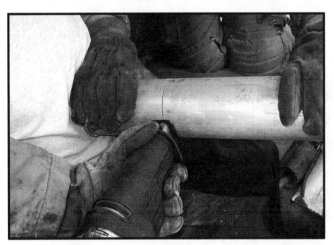

The long blades allow one continuous cut to be made. As the saw cuts through the aluminum, the kerf (gap) may start to close behind the blade and may cause it to bind. If the saw operator sees the kerf starting to close, the other crewmember holding the pipe may be able to make adjustments to keep the kerf open. If the kerf continues to close, the cut can be restarted from the bottom to avoid binding.

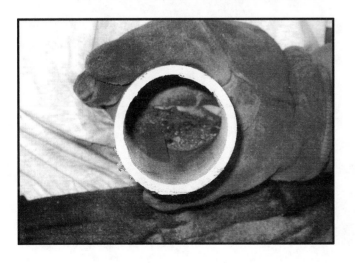

The end result is a clean cut that does not change the shape of the pipe.

Procedure 5
Cutting Large Pipes with a Whizzer

The whizzer may seem like a good tool for an aluminum pipe because it won't produce the sparks that occur when cutting steel. Unlike the reciprocating saw, the whizzer only cuts through one area at a time, making it difficult to end the cut at the same point at which it started.

Without any type of marks to keep the cut straight, it is easy for the cut to move out of alignment.

To complete the cut, the whizzer will have to connect the two ends of the cut.

Crash-Related Impalement

Procedure 6
Cutting Large Pipes with a Hydraulic Cutter

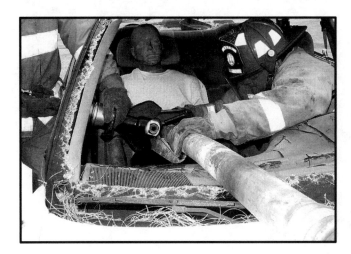

The ability to cut large pipes will be based on the power of the cutter being used.

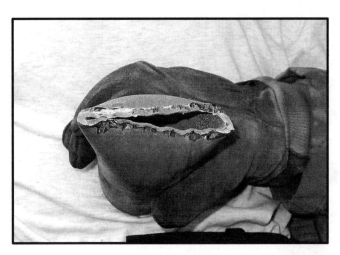

When cutting around hollow objects, cutters will first crush the pipe then sever it with the blade's teeth. The end result is a pipe that is flatter and wider.

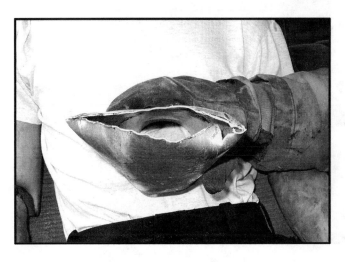

This change in shape may or may not be a problem for the surgeons in the operating room, but, if given a choice between a tool that changes the shape of the object and one that does not, common sense would support using the tool that does not change the shape.

Vehicle Extrication: A Practical Guide

Procedure 7
Cutting Square Stock with a Hydraulic Cutter

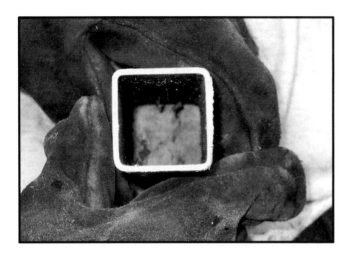

Square stock cut with a reciprocating saw produces clean edges and no deformity.

If cutters are used, the material is placed in the notch for maximum cutting power and minimal movement while being cut.

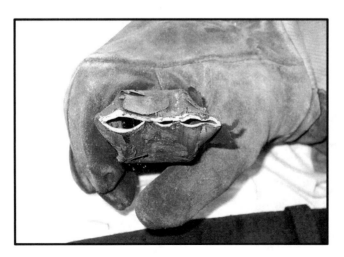

Like the round pipes the square stock flattened out, became wider, and developed some sharp edges.

Crash-Related Impalement

Procedure 8
Cutting Wood with a Reciprocating Saw

When cutting wood, a reciprocating saw with a wood cutting or demolition blade will provide the cleanest cut.

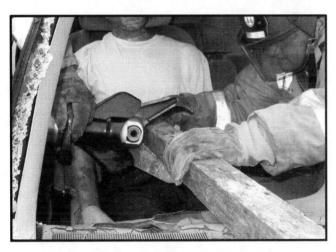

If a hydraulic cutter must be used, it should be positioned so the wood is as deep in the jaws of the cutter as possible.

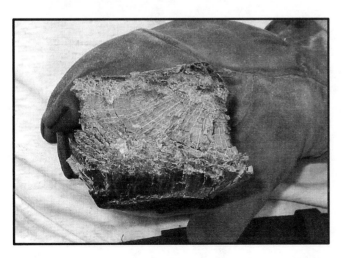

Initially, the cutters will crush the wood and then the teeth will cut it.

·8·

Entrapment beneath and between Vehicles

The success of an extrication operation has always been based on knowledge, skill, appropriate selection of tools, and speed. Speed, while very important, is usually considered secondary to careful handling of the victim and cautious displacement of the metal that is trapping the victim. However, in situations when a victim is trapped beneath a vehicle, the speed of the operation often becomes more important than in other operations. If a victim's chest, abdomen, neck, or head are beneath a vehicle, the weight of the vehicle can make it impossible for the victim to breathe, which can lead to suffocation. In these cases, the ability to size up the situation and act very quickly is essential for victim survival.

While lifting is simple, prepping the vehicle for lifting, selection of lift points, and tool choice require practice, teamwork, and strong leadership. To act quickly, the officer and crew must first determine the manner in which the victim is trapped beneath the vehicle. To put this in perspective, consider the following examples of real-world lifting situations:

1. Unaware of an elderly family member walking behind her vehicle, a driver backed over him. Unaware that she hit someone, the driver continued to back up until the vehicle stopped moving. The elderly man was wedged between the vehicle and the ground with his bathrobe wrapped tightly around the drive shaft.

2. A driver lost control of her vehicle on an interstate highway, entered the median, and rolled. One of the occupants, who was unbelted, was partially ejected through the sunroof as the vehicle came to rest on its roof. With the vehicle upside down, the occupant was lying on her back on the interior of the roof. Her head was through the sunroof, resulting in the vehicle resting on her face, severely compromising her breathing.

3. A pedestrian was walking on the sidewalk when a vehicle left the street, striking the pedestrian. As the car continued to move forward, the victim was run over and trapped beneath the vehicle.

4. An occupant was completely ejected from a vehicle with the vehicle coming to rest upside down on his arm. The victim had no other apparent injuries.

5. A head-on collision caused one vehicle to leave the highway and come to rest in a muddy swale. A size-up of the scene revealed that the driver had been killed, and there was an empty child car seat and infant supplies in the rear of the badly damaged vehicle. A search of the area came up negative, but the muddy area beneath the car was suspect.

While all of the above situations required some type of lifting, each incident required a different approach because of the simplicity or complexity of the situation.

In the case of the case of the elderly man whose bathrobe was wrapped around the drive shaft, the initial priority was to remove the weight of the vehicle from the victim so he could breathe. Another concern was to turn off the vehicle's engine to prevent the victim from being burned by the hot exhaust pipe. After cribbing the vehicle, one side of the vehicle was lifted revealing the drive shaft entanglement. The robe was cut off the drive shaft with a folding knife that was carried by one of the firefighters on the scene.

In the case of the partial ejection through the sunroof, speed was again the critical component in a successful outcome. The soft, sandy soil required the use of cribbing to provide a firm surface while two spreaders were used to lift the vehicle. Without the use of the cribbing as a pad, the spreaders would have been useless and a waste of time.

When the pedestrian was struck and trapped beneath the vehicle, not only was the victim's breathing a concern, but without the ability to move, the hot road surface could have led to serious burns and pain.

In the fourth situation, with the vehicle resting only on the victim's arm, the initial plan was to use high-pressure air bags to perform the lift. One of the firefighters on the scene was quick to notice that the vehicle was lightweight and, based on its position, could be easily rolled off the victim's arm manually by crew members. This was a simple, fast approach based on sound risk/benefit analysis and experience. This procedure will not work in all cases but was certainly the most practical approach in this case.

Attempts to lift in mud or other loose earth may result in failure as the earth is compressed and moves downward, instead of the vehicle moving upward. To overcome the mud problem in the last scenario, a come-along was attached between a tree that was at the top of the swale and the far-side "B" post of the vehicle. By

Entrapment Beneath and Between Vehicles

extending the chains across the roof to the opposite side of the car, the near-side wheels served as a fulcrum, which allowed the far-side wheels to lift off the ground.

When the officer is faced with a person trapped beneath a vehicle, the size-up and subsequent development of tactics should include consideration of the following issues:

- Victim status. The victim's respiratory status is a critical concern and should be evaluated early in the size-up. Ideally, EMS personnel can provide information about the victim's medical status and can start treatment while the victim is still trapped beneath the vehicle.

- Entanglement or impalement. While it may be difficult to see any source of impalement, any clue about entanglement or impalement will help crew members to prepare for dealing with the problem and assist in proper tool selection.

- Best lift point. When a vehicle is still on all four tires, usually, the sides of the vehicle will provide better lift points than the front or rear of the vehicle. An important consideration is the position and location of the victim beneath the vehicle. If the victim's head is near the driver's side of the vehicle and the feet near the passenger side, it would be best to lift from the driver's side. The advantage of lifting from the area closest to the victim's head is twofold: first, by lifting closest to the head there is a greater chance of relieving pressure on the chest, head, and neck. Second, when the victim is dragged in-line from beneath the vehicle, the lower portion of the body is smaller making it easier to move through tight areas.

- Best tool for the job. Experience is the best teacher when selecting the most effective tool for a lift, and the most practical place to get this type of experience is on the drill ground. Spreaders can provide a quick lift but will not provide the most stable lift while air bags provide very good stability but require a little more training to obtain the best results. On the other hand, a 5-ft pry bar with a fulcrum point near the vehicle may be all that is needed in some situations. The best way to learn about lifting is to put a manikin under a vehicle and try different tools and procedures. Even if the victim can't be completely freed with the tools on the first arriving unit, initial actions by the crew may allow the victim to breathe while waiting for other units to arrive with more tools.

- Is digging out an option? If soft earth and light entrapment is encountered, digging the victim out may a reasonable option.

General Procedures

When performing lifting procedures, in most situations the following tasks should be included in the operation:

- Chock the tires on the side of the vehicle not being lifted. This will help prevent the vehicle from rolling away as the vehicle is lifted. This should be done even if the vehicle is on a level surface because, as the vehicle is lifted, the lifting tool may exert pressure resulting in the vehicle being pushed forward or rearward.

- Set the parking brake. This is extra insurance against the vehicle rolling away. The person setting the brake shouldn't sit in the vehicle to set the brake as this extra weight will be transferred to the victim.

A good size-up will reveal if the victim is truly trapped or simply beneath the vehicle.

The parking brake is set to reduce the chance of the vehicle rolling away.

The ignition is turned off.

The transmission is moved to "Park" if not already in that position.

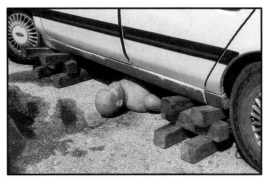

Cribbing is put in place to prevent any further downward movement of the vehicle.

The tires, especially on the non-lifting side, are are chocked to prevent movement.

Fig. 8–1 Lifting basics—steps before the lift begins

Entrapment Beneath and Between Vehicles

- Turn off the engine and remove the key. A hot engine, transmission, or exhaust system can cause disabling burns if the victim is trapped against these components.

- After the engine is turned off, put the transmission in park if it's an automatic, or put it in gear if it's a standard transmission. Again, the crew members should avoid entering the vehicle to avoid transferring weight to the victim.

In soft soils, digging the victim out may be an option when lifting tools aren't available.

Even if complete removal isn't possible, digging may relieve pressure on the victim's chest.

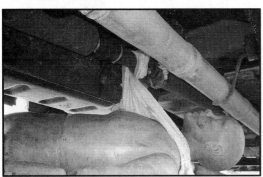

A good size-up will identify any entanglement of body parts or clothes.

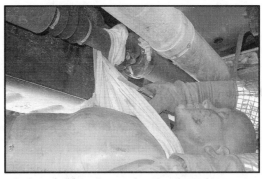

A sharp knife or heavy-duty bandage shears can be used for cutting away clothing.

There aren't many good lifting points at the front or rear of automobiles.

Lifting from the side will provide the best lift for the front and rear of the vehicle.

Fig. 8–2 Lifting basics

- Crib the opposite side of the vehicle to prevent downward movement during the lift. As a vehicle is lifted from the side, the two tires on the opposite side of the vehicle serve as fulcrum points. This situation, along with movement of the suspension system, may actually cause the opposite side of vehicle to move downward on the victim. By cribbing under the non-lifting side of the vehicle, this downward movement can be eliminated enhancing the lift operation.

Hand tools and wedges

Wedges can provide rescuers with a quick way to lift a lightweight vehicle a couple of inches without the use of any complex tools. This procedure can require a lot of cribbing, necessitating a careful size-up by the officer, and an accurate prediction of the amount to cribbing needed. If an incorrect estimation of cribbing needs is made, the crew may spend time setting up the operation only to find they don't have the cribbing required to complete the job. The procedure is simple. To obtain a couple inches of lift, drive a wood wedge into a gap between the vehicle and a box crib. While effective in some circumstances, this procedure should not be considered a first or even second choice when there are other options available.

Pry bars

Pry bars have built-in fulcrum points that provide a tremendous mechanical advantage when lifting an inch or so, and when the space between the ground and the vehicle is less than an inch. Unfortunately, this built-in fulcrum point isn't very useful when trying to lift a vehicle off of a victim because of the typical amount of space between the underside of the vehicle and the ground. Pry bars can still be used to good advantage when a large piece of cribbing (6x6), a step chock, or box cribbing are used as fulcrum points. When selecting tools, if there is a choice of two pry bars of different lengths, the longest should be selected, or if personnel are available, both should be used simultaneously.

Following are important points to remember when using pry bars:

1. The longer the bar, the greater potential mechanical advantage.

2. Placement of the fulcrum point (cribbing, etc.) will determine the amount of mechanical advantage created.

3. When pressing down on the pry bar, the closer the operator's hands are placed to the end of the bar, the greater the mechanical advantage.

4. Heavier tool operators can exert more downward force on the bar, which means greater lifting force. This is a good time to put the big guys on the tool.

5. As quickly as an object is lifted, it can be dropped if the fulcrum or vehicle shifts, or if the tool operator loses control of the bar. Cribbing as the lift is performed is very important when using this tool.

Hi-Lift jack

The Hi-Lift jack can provide a very quick lift if there is a sturdy, accessible lifting point available. As with other tools used in lifting operations, as

Entrapment Beneath and Between Vehicles

the tool lifts and the vehicle is tilted, the position relationship between the vehicle and the jack may change, requiring constant assessment and adjustment as needed.

The following are important points to understand when using a Hi-Lift jack:

- The Hi-Lift instruction manual should be consulted for specific instructions on its use and capacities.

- The amount of contact between the nose (lifting surface) of the jack and the vehicle may change as the vehicle rises, necessitating careful monitoring.

- Lifting from the side of a vehicle will produce a more pronounced movement of the vehicle away from the jack than when a vehicle is lifted from the front or rear. This is caused by the short distance between the jack and the tires on the opposite side of the vehicle. This means that it is likely that the jack will move from a vertical position to more of a leaning position. This can translate into decreasing stability that should be monitored.

- When it becomes necessary to lower the jack, the tool operator should bring the handle to the upright position and then clip it against the steel standard before moving the reversing lever to the "down" position. When lowering the jack, the tool operator should have a firm grip on the handle with both hands. As the load is lowered, the upward force on the handle is comparable to the force that is required to push the handle downward during the lifting part of the operation. Lack of handle control can result in undesirable rapid downward movement of the vehicle and an equally undesirable rapid upward movement of the handle.

Come-along

The come-along may be the tool of choice when operating in muddy or soft conditions that cause other lifting tools to sink into the earth. To use the come-along to lift, there must be a strong anchor point off the non-lifting side of the vehicle. Instead of lifting from the bottom up, as with other tools, this procedure pulls across the top of the vehicle toward the opposite side, causing a rolling type movement. Unlike other lifting operations, box cribbing on the side of the vehicle opposite the victim won't be very effective. This is because as the come-along is operated, the vehicle will have a lateral, or side load placed on it, which would cause a box crib to slide apart.

Step chocks or wedges may work, but after analysis no cribbing may be the best choice for the non-lifting side of the vehicle. As the vehicle is lifted, cribbing is required on the lifting side of the vehicle. Throughout the operation, the company officer and the crew member operating the come-along should monitor the anchor for any sign of movement or indication of possible failure.

Manually operated spreaders

Manually operated spreaders don't have the large opening distance that heavy hydraulic spreaders have, which translates to less lifting height. To maximize the potential lift of manually operated spreaders, the gap between the ground and the object being lifted should be as small as possible (see Fig. 8–5). Following are a couple of steps that can maximize the lifting ability of this type of spreader:

- Place the spreaders in the area that is the smallest distance between the vehicle being lifted and the ground if the terrain is uneven.

- Use a single, large piece of cribbing (such as a 6x6) to close the gap and make the lift more effective.

High-pressure air bags

When the lift shifts from simple to complex, the high-pressure air bags provide the best method of lifting. The advantages of the air bags over other tools include the following:

- The shape and size of air bags result in a very stable lift.

- Usually a minimum of two bags is available, providing the opportunity to perform a high lift by stacking the bags or a heavy lift by placing the bags side by side.

- The height of the vehicle can be increased or decreased easily without any sudden movement or removal of equipment.

- Tipping over isn't usually a concern.

Following are important points to understand when using air bags to lift:

- If available, two air bags should be used for each lift point. When lifting cars and small pickup trucks, the important factor is the maximum height available from the air bag. The weight capacity of the air bag is seldom an issue, as even full size pickup trucks usually weigh only about three to four tons.

- The air bag lifting points should be carefully selected to provide adequate room for the bags, cribbing, and removal of the victim. If possible, two lifting points should be utilized, but this may be impractical on most lifts because of space limitations.

- Lifting points should be evaluated for weakness caused by rust, damage, or vehicle design. Exhaust components should be located and avoided to prevent burn damage and failure of the air bags. Fuel tanks and sharp objects on the underside of the vehicle should also be avoided.

- On soft ground, a ground pad should be used to create a firm base and to reduce the amount the bags sink into the earth.

- A solid box crib should be built as a base for the air bags if necessary. The purpose of the box crib is to move the starting position of the air bags as close to the bottom of the vehicle as possible. This will result in additional inches of lifting potential, if needed.

- When the lift begins, the air bag is inflated until it makes light contact with the vehicle. A crew member then checks the positioning for any problems and corrects them as needed.

- After the position of the air bags is checked, the lift can be continued. If the controller operator inflates the air bags too quickly, the crew members who are building the cribbing will have a hard time keeping up with the lift.

- Cribbing supplies should be positioned close to the area where they will be used to reduce delays.

Heavy hydraulic spreaders

Heavy hydraulic spreaders were designed for prying doors open, but can be used for lifting operations as well. When used for lifting, the tool operator has a few more concerns that must be addressed to maintain a safe operation. Recognizing that spreaders are being used for lifting, manufacturers have developed attachments that improve the operation.

Following are important points to understand when using spreaders to lift:

- Hydraulic spreaders are prone to tilting or falling over as the height of the lift increases. This problem requires the tool operator to be aware of not only the position of the spreaders during the lift but also to anticipate at what point the tool may become unstable and fall over.

- As the lift proceeds, the contact area between the lower spreader tip and the ground will decrease, creating two problems. The first problem is the increased likelihood of the spreaders sinking into the ground. This is caused by the weight of the vehicle being concentrated into an area that is decreasing in size during the lift. The second problem is the decrease in stability as the contact area between the tip and ground decreases. This problem becomes more severe as the height of the lift increases.

- With experience, the tool operator will see that positioning the lower tip inward slightly (toward the opposite side) will be helpful in higher lifts. This is because as the vehicle is tilted, or rolled upward, the side of the vehicle being lifted will move up and away from the operator. By starting the lift with the lower spreader tip positioned a little more inward than the upper tip at the beginning of the lift, the tips will align vertically as the lift progresses.

- Spreader attachments designed for lifting can dramatically improve the procedure by providing a larger base that contacts the ground, providing greater stability.

- If a spreader attachment designed to serve as a ground pad is available, it's placed on the spreader arm that will placed against the ground.

- The tool operator should examine the lift point on the vehicle to determine its strength. Rust and damage may cause the metal to tear during the lift, resulting in an unanticipated drop of the vehicle.

- The crew member operating the spreader should recognize that spreaders can raise a vehicle quickly, making it difficult for the crew members building the cribbing to keep up.

Depending on the circumstances, the company officer may consider performing the lift quickly and pulling the victim out without the use of any cribbing. Before using this approach, it's important to first consider the outcome if the spreaders were to tip over or the vehicle were to shift. Additionally, if a simple lifting operation turns into a complex one, requiring unanticipated disentanglement or the repositioning of the lifting tool, the crew may find that they can't continue until cribbing becomes available and is set up. As always, a good size-up and plan will lead to the best results.

Procedures

Vehicle Extrication: A Practical Guide

Procedure 1
Lifting a Vehicle with Wedges

A box crib and a few wedges can be used to quickly achieve a few inches of lift while other tools are being set up or brought to the scene. While a few inches may not seem like a lot, it may provide enough space for the victim's chest to expand, allowing the victim to breathe while still trapped beneath a vehicle.

To prepare the lift, it is best to build a solid box crib as the base for the wedges. The gap between the bottom of the vehicle and the cribbing is closed with cribbing as much as possible before inserting the wedge. Once in place, the wedge is driven into the gap with a sledge hammer or flat head ax. As the wedge is driven in, the body of the vehicle will move upward on its suspension.

Obstructions and gaps beneath the vehicle can complicate the procedure if they are not addressed at the beginning of the operation before the cribbing is put in place. Setting up the cribbing to avoid obstructions can result in a smoother, higher lift.

Entrapment Beneath and Between Vehicles

Procedure 2
Lifting with a Pry Bar and Cribbing

The ability to lift manually is a basic skill all firefighters should be capable of performing. This simple class one lever system uses a box crib as a fulcrum and a 5-ft pry bar as a lever. The top piece of cribbing is prevented from sliding out until the vertical load holds it in place. The vertical load is created by the pry bar as the crewmember presses down.

This lifting method may not be adequate in many cases but may provide enough space to allow a trapped victim to breathe. The set-up is quick and simple. In some cases, the victim who is only slightly trapped may be freed with the few inches of lift achieved with the pry bar. As always, cribbing should be utilized to support the load as it is lifted.

There is a direct relationship between the weight of the crewmember on the pry bar and the amount of weight lifted at the other end. The heavier the firefighter, the greater the weight that can be lifted. If two pry bars are available, they can be used to double the amount of weight lifted.

Procedure 3
Lifting a Vehicle with Manual Hydraulic Tools

Some heavy rescues carry floor jacks for lifting operations for the simple reason that the sole purpose of a floor jack is to lift. A floor jack has a low profile that allows it to be slid deeply under a vehicle and a lifting pad designed for the chassis of a vehicle. When positioning a floor jack for a lift, the lifting point on the vehicle should be examined for rust and overall sturdiness.

Once in position, the handle is rotated clockwise to close the valve. To lift, the handle is pumped up and down. Each downward stroke of the handle pumps hydraulic fluid and forces the lifting pad upward. To lower, the handle is turned counterclockwise.

A manually operated spreader can be used for lifting like a power hydraulic spreader but with less lifting height. Manually operated spreaders usually have a smaller opening distance between the tips, which equates to a smaller lifting distance. To minimize the lack of lifting height, a large piece of cribbing can be used to elevate the lower spreader arm.

Entrapment Beneath and Between Vehicles

Procedure 4
Lifting with High-pressure Air Bags

Air bags can provide the most stable and powerful lifting options for freeing a victim trapped beneath a vehicle. If air bags are used, the crewmembers should recognize that the work area is going to be crowded and plan for the space required for the air bags, cribbing, and victim removal.

With a limited amount of room available, step chocks provide a good option for cribbing. They are stable, have a reasonably sized base and a narrow profile that can be advantageous in a crowded work area. Additionally, the step chocks can be slid into position as the lift progresses. The downside to using step chocks becomes apparent when the maximum height of the chock is reached and additional cribbing can not be easily added.

When the planned lifting height exceeds the height of the step chock, box cribbing provides the best alternative. As the height increases, additional members can be added to the top of the box crib. Wedges can be used to fill the gaps that are too small for the full-size cribbing members. As the box crib is built and adjusted, crewmembers should remember not to obstruct the area that will be used for victim removal.

Procedure 5
Lifting with High-pressure Air Bags

When the needed lifting height exceeds the height available from one bag, two bags can be stacked. Even when it appears that only one bag will be needed, it's best to use two stacked bags in case the required lifting height is underestimated, or the bags sink into the soil. Additionally, two bags provide a backup option in case of equipment failure that causes one of the bags to become unusable. When using two air bags, most firefighters prefer to inflate the bottom bag first, followed by the top bag. Three bags aren't used unless part of a system specifically designed for that configuration.

If only one air bag is used, it's best to decrease the distance between the underside of the vehicle and the ground by building a box crib to be used as a platform for the air bag. The platform allows the lifting potential of the air bag to be used for actual lifting instead of merely filling the void between the vehicle and the ground. A solid box crib is the best configuration if enough cribbing is available. If not, the top layer should be solid to give the air bag something to push against as it's inflated.

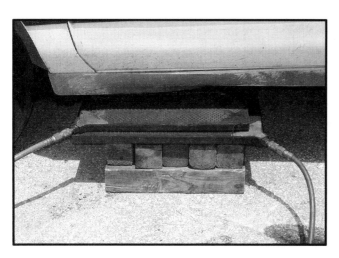

The best configuration is when two air bags are used and a solid box crib is built as a platform. This allows the height of both air bags to be used while being supported by a solid, stable platform.

Entrapment Beneath and Between Vehicles

Procedure 6
Lifting with Hydraulic Spreaders

Lifting the weight of a vehicle is well within the capacity of hydraulic spreaders and is one of the common ways to make a quick lift. If possible, the lower tip is positioned so that it is further beneath the vehicle than the top tip. This positioning will help keep the tips vertically aligned once the vehicle starts moving up and away from the tool operator. This movement is caused by the tires on the far side of the vehicle becoming pivot or fulcrum points for the vehicle.

The ability to position the bottom tip further beneath the vehicle will be dependent on the height of the near side of the vehicle. If the height of the vehicle is equal to or less than the height of the spreader tool as it lies on its side, it will be difficult to maintain the desired positioning. As the lift is being prepared, cribbing should be assembled to support the vehicle as it is being lifted.

As the vehicle moves upward, cribbing is put in place to support the load. As the lift proceeds higher, the spreaders become less stable. If the vehicle shifts even a little, the spreaders will want to tip over sideways. Unfortunately, most attempts to prevent spreaders from tipping over once they have started are unsuccessful. To minimize the chance of tipping, the spreaders should only lift as far as needed to remove the victim from beneath the vehicle.

Vehicle Extrication: A Practical Guide

Procedure 7
Lifting with Hydraulic Spreaders

When planning a lifting operation in any soft soil environment, the crew should plan on the spreaders sinking into the ground. While the ground may be dry and firm enough to walk on, when the spreading force of the tips is concentrated in the very small area making contact with the ground, the spreaders will often sink into the earth. The best way to deal with the problem is to plan on it happening and taking steps to prevent it.

To overcome the concentrated force of the tips, the load must be distributed over a larger area. One simple way to spread the load is to place a piece of cribbing beneath the spreader tips. Once the spreaders start to lift, it will become obvious if the load has been distributed over a large enough area. If the single piece of cribbing sinks into the earth, two or three additional pieces can be placed beneath the single piece. This will allow the load on the single piece to be redistributed over the surface area of three pieces.

Special attachments are made for spreaders that not only distribute the load, but ones that also provide a pivot point between the lower spreader arm and the ground. Without the pivot point, the attachment pad would not stay on the ground. If this pad does not provide enough size to support the load, it can be placed on a larger object such cribbing or an outrigger pad.

Entrapment Beneath and Between Vehicles

Procedure 8
Lifting on a Hard Surface with a Hi-Lift Jack

As with all lifting operations, cribbing is built to the underside of the vehicle before the lifting operation begins. After the jack is moved to the area where it will be used, the reversing lever (operating lever) is moved to the "up" position. This can be done with the edge of a boot or by hand.

The jack is slid into position. While holding the top of the jack firmly with one hand, the handle is unclipped from the steel standard and pivoted outward.

Still holding the top of the jack, the handle is pulled upward, sliding the lifting mechanism up the steel standard until the nose of the lifting mechanism makes contact with the vehicle. The lifting point on the vehicle should be solid and flat. Rounded lifting points can slip off the nose of the jack.

Vehicle Extrication: A Practical Guide

Procedure 8 *(continued)*
Lifting on a Hard Surface with a Hi-Lift Jack

During the first downward motion of the jack handle, the tool operator may want to hold the top of the jack in place until the nose of the jack has firm contact with the vehicle. Once good contact is made, the operator should maintain a good grip on the handle with both hands. As the lift is made, cribbing is built to support the vehicle in at least two spots.

During the lift, the contact between the jack nose and the vehicle should be monitored for movement or slippage.

As the jack lifts the vehicle from the side, the side of the vehicle will move up and away from the jack. As this occurs, the jack will move from the vertical position to a tilted position. As the jack tilts, it becomes less stable, increasing the chance of it slipping off the vehicle. As with all lifting operations, the vehicle should only be lifted to the height needed to remove the victim.

Entrapment Beneath and Between Vehicles

Procedure 9
Lifting on a Soft Surface with a Hi-Lift Jack

The Hi-Lift jack is well suited for lifting on both hard and soft surfaces. Its relatively light weight makes it a good choice when rescue vehicles and heavy equipment cannot get close to the incident location.

The standard foot for the Hi-Lift jack provides about 28 in. of surface area to support the load. This amount of surface may be enough to support the entire load on some soils but may need additional support on others.

The Hi-Lift jack is used primarily for lifting vehicles in off-road situations and has accessories designed to facilitate this use. The Hi-Lift off-road base provides a large base that the jack sits on, reducing the likelihood of the jack sinking into the soil. The off-road base is made of plastic and is lightweight.

Procedure 9 *(continued)*
Lifting on a Soft Surface with a Hi-Lift Jack

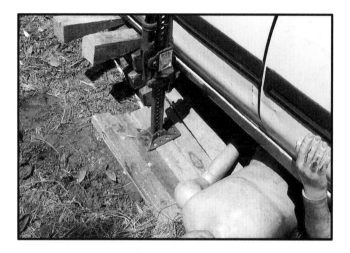

If cribbing is available, a few pieces can be placed beneath the foot of the jack to improve the load distribution in soft soils. When deciding how much cribbing to use for the base, the tool operator should remember that cribbing will also be needed to support the load as the lift progresses. For every few inches of lift achieved, several pieces of cribbing will be needed for the box crib that will support the vehicle in case the load shifts or falls.

Makeshift aids can also be used to help support the jack and may be found around a crash scene or in one of the vehicles. The key is to use something that has a large surface area, is fairly sturdy, and is unlikely to fail suddenly or without warning. A spare tire and wheel assembly can provide a quick and sturdy base. The wheel can be flipped one way or the other to provide different platform heights.

In a perfect world, all the required tools would be available to perform any rescue procedure. In the real world, tough decisions must be made when the best tools are not available for use. In these situations, crewmembers must make decisions based on their experience and training. This wheel cover was not designed for support loads but will support them anyway. The ability to evaluate and adapt are valuable skills at the rescue scene.

Entrapment Beneath and Between Vehicles

Procedure 10
Lifting with a Come-along in Soft Soil

A come-along can be used to lift one side of a lightweight vehicle when there is a strong anchor point on the opposite side of the vehicle from where the victim is trapped. This procedure may be the best choice when the ground is soft providing no support for lifting tools. If the tires are not dug into the ground, wedges or step chocks can be used on the non-lifting side to prevent any sideways sliding of the vehicle. This procedure does not work well on paved surfaces because the tires will slide sideways instead of lifting as the come-along is operated.

To provide lift on the driver's side of the vehicle, the tires on the passenger side of the vehicle will be used as a fulcrum. To create the lifting motion, a chain is attached to the driver's side "B" post. To keep the chain from falling off the roof, the master link and chain are pushed to the passenger side of the roof.

With the excess chain and the master link positioned on the passenger side of the roof, the chain stays in place. When the sling hook is equipped with a latch, it further reduces the chance of the chain slipping out of the hook. The chain should be moved high on the post to maximize the effectiveness of the pull. When snugged up, the chain should be above the shoulder harness hardware.

Vehicle Extrication: A Practical Guide

Procedure 10 *(continued)*
Lifting with a Come-along in Soft Soil

The come-along is attached to the master link that is already on the passenger side of the roof. The crewmember operating the come-along will have to estimate the amount of cable that will be needed to reach the anchor point. With an accurate distance estimate, the crewmember can decide if the come-along can be used in the 2:1 mechanical advantage configuration (doubled back) or if all the cable on the come-along will be needed to reach the anchor point.

The anchor chain is attached to the anchor point at a height that will allow the come-along to be operated effectively.

Fastening the anchor chain can be accomplished by hooking to a solid anchor point, by choking as shown on the tree, or by wrapping the chain around the anchor and using the chain shortener. In this procedure, choking was used because there was no manufactured attachment point to attach the hook to, and using the chain shortener would have greatly reduced the reach of the anchor chain.

Entrapment Beneath and Between Vehicles

Procedure 10 *(continued)*
Lifting with a Come-along in Soft Soil

Once the crewmember has attached the come-along to the master link, the come-along is backed out to the anchor rigger. If proper distance estimates were made, the come-along will reach the anchor chain with a couple wraps left on the come-along drum. If the cable comes up short in the 2:1 mechanical advantage configuration, the come-along can be switched to a 1:1 ma or another chain can be used to extend the reach.

With the system completed, the officer directs the come-along operator to take up the slack cable. The crewmembers are then directed to check their rigging for any problems that may have occurred when the slack was taken out of the cable. For longer 1:1 pulls, a coat or other object can be placed on the cable to provide a little weight in the unlikely chance the cable were to break. If the cable were to break, the coat would cause the cable to fall to the ground instead of springing back toward the operator or vehicle.

After the system is checked, the officer directs the crewmember to operate the come-along. If the vehicle starts to slide, the operation should be stopped immediately and corrective action taken to stop any further sliding of the vehicle. When the vehicle starts to lift off the ground, cribbing is added and adjusted to fill the gaps. The possibility of sideways sliding makes this procedure less reliable than other procedures that should be considered first.

Vehicle Extrication: A Practical Guide

Procedure 11
Lifting an Overturned Vehicle with a Hi-Lift Jack

A Hi-Lift jack can be used to lift one side of an overturned vehicle in a couple of ways. One method is to use the jack in the standard manner with the nose of the jack placed in the window opening. If the shape of the metal prevents the jack from reaching the window opening, another option involves the use of a chain.

To create lift on one side of the vehicle, a chain is run through the passenger compartment and attached to the opposite side of the vehicle. The chain is then attached to a solid frame member.

Another option is to attach the chain to the opposite side roof post. In the photo, the chain is attached to the opposite side "B" post. If a standard 12-ft chain assembly is used, it is best to choke the "B" post. If the chain is equipped with a grab hook, the hook can be attached to the chain close to the "B" post to maximize the amount of available chain.

Vehicle Extrication: A Practical Guide

Procedure 11 *(continued)*
Lifting an Overturned Vehicle with a Hi-Lift Jack

After the chain has been attached to the opposite side, the master link is connected to a hook on the Hi-Lift jack and the excessive chain removed with the chain shortener. The operating lever is lifted until it clicks into the "raise" position. The handle is then pulled upward to take out any remaining slack in the chain. The vehicle is cribbed to prevent any downward movement during the lift. Once the slack is removed and cribbing is in place, the system is checked. After the final check, the lift is started.

The jack is operated until the desired height is attained and is stopped before the top of the jack is reached. The lift would also be stopped if any undesirable shifting of the vehicle occurred or if the jack were to become unstable.

This procedure will not provide a lot of lift because so much of the jack height is used in the initial positioning and removal of slack chain. As with all lifting operations, the standard *lift an inch, crib an inch* rule should be used to maintain the vehicle height in case of an unexpected movement of the jack or the vehicle.

Vehicle Extrication: A Practical Guide

Procedure 12
Spreading Two Vehicles with Hydraulic Tools

After a crash, a vehicle may come to rest in a position that results in the pinning of an ejected occupant or pedestrian. If the vehicles trapping the victim can not be moved away from the victim in the normal manner, hydraulic tools can be used to create the space needed to free the victim.

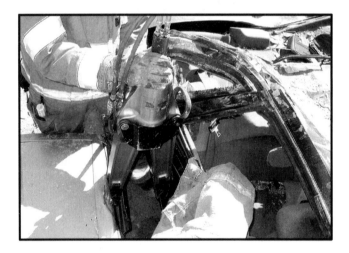

Spreaders are inserted between the vehicles in sturdy areas and opened. If the metal collapses, the spreaders can be repositioned.

The maximum spreading distance is based on the spreading distance of the tool. After one area is spread apart, the tool can be repositioned as necessary.

Entrapment Beneath and Between Vehicles

Procedure 12 *(continued)*
Spreading Two Vehicles with Hydraulic Tools

If more space is needed, a hydraulic ram can be used.

The painted surfaces of the vehicle body may cause the ends of the rams to slip, requiring careful observation to spot the problem before it occurs.

A large ram fully extended should provide adequate room to access and remove a victim.

·9·

Trucks and Tractor Trailers

There are several ways of laying the groundwork for learning truck extrication skills including an overview of the history of trucks, truck design, manufacturing processes, and the statistics about the number of trucks involved in accidents on an annual basis. That information is nice to know but may not provide the stepping-stones needed to transition from automobile extrication skills to truck extrication skills. To provide the necessary learning links, a more meaningful and productive approach will be used, which provides the basic information about truck construction and extrication procedures.

COMMON TERMS

One proven method of learning new information is to build on existing knowledge. One example of this is approach is to start with clarifying the commonly used terms *truck* and *tractor trailer*. Both types of vehicles have powerful engines, are capable of handling heavy loads, and may look similar when in use.

Vehicle Extrication: A Practical Guide

While the terms *truck* and *tractor trailer* are often used interchangeably, the terms *truck* or *straight truck* are used to describe a vehicle that has a chassis and cab designed to have some type of cargo box or machine directly attached to the frame behind the cab. Trucks can be outfitted as garbage trucks, delivery trucks, tow trucks, and many other uses.

Tractors (including farm tractors and highway tractors) are designed to pull a trailer or attachment behind them that can be removed

Straight truck

Conventional tractor with tall sleeper equipped with sleeper access door and windows.

Conventional with flat top sleeper and an add-on fairing on top of the sleeper.

Conventional with sleeper, access door and windows.

Cab forward in the foreground, straight truck in the background.

Cabover with a windowless sleeper and fairing on top of the sleeper.

Fig. 9–1 Truck and tractor styles—a comparison of trucks and tractors that are similar in appearance.

Trucks and Tractor Trailers

and replaced with another. A farm tractor pulls farm equipment, and a highway tractor pulls a large trailer. While trucks and highway tractors are similar in appearance, understanding the distinction between the two is helpful when communicating effectively.

There are a few styles of trucks and tractors including the *conventional*, *cab-over*, and *cab-forward*. The conventional highway tractor is set up like an automobile; the engine is at the front of the vehicle covered by a big hood, and the driver sits behind the front wheels. The cab-over highway tractor can be recognized by its big, boxy shape and its tall flat front, with the cab and the driver positioned over the front wheels. The cab-forward design is being seen more often around town and is commonly used for smaller delivery trucks and sanitation trucks. The description "cab-forward" means the cab is in front of the engine.

While the thought of working on the large, conventional rig used to haul massive loads may seem intimidating, it should be recognized that the size of interior driver and passenger spaces are similar to other vehicles. While at first that may sound impossible, the distance from the seat to the pedals on the floor are basically the same in both vehicles because the people who drive them are basically the same. In fact, anyone who has ridden in the front seat of some of the new fire apparatus would probably say that there's less space in those apparatus than in a full-size pick-up truck.

The similarity between the big highway tractors and automobiles doesn't end there. In fact, like the hybrid car in chapter 3, it's easier to understand the differences between two types of vehicles if the similarities are first recognized. Like an automobile, a truck or tractor has two doors, an engine, a dashboard, and an ignition system. To make them safer to work around, the brakes are engaged, the engine shut down, and the batteries disconnected when possible.

The big rigs have tilt and telescoping steering, just like a luxury sedan, along with adjustable seats that may make it easier to remove the occupant from the vehicle. Most of the extrication procedures are the same, but when operating on trucks, the tool operator must use the tool from a different angle and position. Doors are popped, windshields removed, and dash roll-up procedures are very similar to automobiles, with the addition of a couple of steps.

Crews have to address and adapt to a few differences when performing a truck or tractor extrication. Addressing the differences in the order that they would normally be encountered, the first thing crew members will notice when sizing up a crash will be the overall size of the vehicle and height of the occupant compartment.

Stabilizing a large tractor in a precarious position may be difficult and may require a tremendous amount of cribbing. Even if the cribbing is available, a 3-ft tall box crib probably won't support a tractor as it slides sideways off a wall. Manufactured stabilization equipment, come-alongs, and chains may be needed to keep a truck from moving. Adaptability is important, and the officer should consider real-world solutions to the problem(s) at hand. It's possible that a heavy-duty wrecker or even a piece of heavy equipment may provide the real-world choices for assisting in stabilization of the tractor or truck.

Vehicle Extrication: A Practical Guide

Fig. 9–2 Heavy duty wreckers rated for heavy loads may be needed for real-world stabilization problems.

Fig. 9–3 Utilization and adaptability of available resources may solve some of the real-world stabilization problems that can occur.

Handling heavy tools above shoulder height is difficult, making the use of a ladder a necessity in some cases.

POPPING A DOOR WITH HYDRAULIC SPREADERS

If a door needs to be popped open, the spreader operator will find it necessary to approach the latch from below instead of above as is most common when working on automobiles. If it's necessary to remove a door at the hinges after popping it open, two types of hinges will be encountered: the standard automobile hinge and the piano hinge. A piano hinge is a continuous lightweight hinge that spans a distance of a couple of feet. The way a piano hinge is attached makes it difficult to spread or cut apart using the standard automobile methods. An air chisel can be used to sever the entire hinge, or if hydraulic tools are used, the cutters can be used to make a cut through a portion of the hinge that can then be used as a spreader purchase point.

These doors are large, quite heavy, and difficult to control when removing them at the hinge. To

Fig. 9–4 and 9–5 A rope can be attached to the door that is to be removed, thrown over the roof to the other side so a crewmember can hold the door in position for better control and enhanced safety.

388

enhance safety and control of the door, a rope can be attached to the door that is to be removed and then thrown over the roof to the other side of the vehicle. The crew member can hold the rope and door in position while it's being cut and then lower the door after the cuts are completed.

With the door out of the way, the ignition switch and parking brake activators can be located if access to them was denied prior to popping the door. Unlike automobiles that use a hand lever or foot lever to set the parking brake, most large trucks are equipped with air brakes and a button that locks the brakes. These buttons are usually yellow and easily located within the driver's reach.

Fig. 9–7 A truck's laminated windshield glass may be difficult to reach and may require the use of a ladder.

Fig. 9–6 This common style of parking brake control is often yellow in color and is located within easy reach of the driver.

Removing a laminated glass windshield on a truck can be a little more challenging than an automobile because of the height.

After putting a ladder in place, the glass can removed easily by pulling the rubber gasket and the glass out of the windshield frame in a few big pieces. Without a ladder, it may be possible to hook a broken windshield gasket with a pike pole and pull it out of the frame. In the unlikely chance that the windshield that needs to be removed has remained intact, the rubber bead in the gasket can be pried out and then pulled to release the gasket and glass. Another option is to split the gasket with a knife, which in turn will release the glass. Once removed, handling the windshield can be awkward and may require assistance from another crew member.

Fig. 9–8 A heavy duty knife can be used to cut and separate the portion of the windshield gasket that holds the glass in place. Once severed, the gasket is pulled away and the windshield pushed outward, toward the front of the vehicle.

If a dash roll-up or dash lift is required on a conventional cab, the size and mass of the vehicle may seem intimidating at first but flipping the hood open will reveal the true size of the problem. With the hood out of the way, access to the firewall and dashboard areas are enhanced, making them easier to size up and work on. Removal of the hood will also reveal the structures that provide strength to the firewall and cab so plans can be formulated on how to defeat them.

Performing a Dash Roll-Up

When performing a dash roll-up or dash lift on an automobile, the dashboard and firewall are pushed upward or forward out of the occupant compartment. To eliminate resistance and achieve an easier push or lift, the structural support members in front of the firewall that would resist the movement (typically the top rail) are located and severed. The same is true on the trucks. By removing the hood, the obvious (and not so obvious) structural members can be located, severed, or otherwise removed as resistance to the movement of the firewall and dashboard. Design and construction of the vehicles will be different, but with careful observation and identification of the parts that would prevent the desired movement, the necessary cuts can be planned and made.

In addition to removing the strength of the cab in front of the firewall, the standard relief cuts are made at the forward lower corner of the door. As with automobiles, severing this area allows the hinge pillar to separate from the floorboard thereby removing resistance to the forward or upward movement of the dashboard.

When working on cab-overs, as the relief cut in the hinge pillar is planned, the crew may notice that the cut can easily be extended across the front of the cab. Any additional cutting will certainly help in weakening the structure, making it easier to move the dashboard. In fact, if the cut is extended horizontally to the center of the cab, a vertical cut can be added to the evolution resulting in total removal of the dashboard.

Successfully removing the dashboard on a cab-over is centered on severing the main support that runs from door to door in the dashboard. This may require multiple cuts from the front of the dashboard and cab, above the dashboard and from the interior. The severing of this large structural member can be accomplished with a reciprocating saw, an air chisel, or large hydraulic cutters. Even if the supporting member can't be completely removed, any work done on it will make a dash roll-up or lift easier.

Removing, rolling, or lifting the dashboard may be required to free the occupants in the front of the vehicle, but other procedures may also be needed when the vehicle is equipped with a sleeper compartment behind the cab. Not all big trucks or tractors have sleepers; large vehicles that are used around town on an 8:00 AM to 5:00 PM schedule are known as daycabs. These vehicles don't have sleepers because they're only used during normal business hours. But when the tractor is used for hauling loads out of town, sleepers provide the driver or driving team a place to sleep during legs of the trip. Many of the sleepers are simple and large enough to accommodate only a mattress. As the sleeper becomes larger, they can include bunk beds, refrigerators, sofas, and closets.

Trucks and Tractor Trailers

When involved in a serious crash that incapacitates the driver, tractors with sleepers require what can be best compared to a primary search for occupants. While one member of the driving team is operating the vehicle, the other driver or children traveling with their parents could be sleeping in the sleeper. A search of the sleeper is an important step in a good size-up.

If a victim is found in the sleeper and can't be removed from the normal entry point, crew members will have to remove the victim via an access door in the sleeper or by removing part of the sleeper wall or roof.

The good news about sleepers is that in the interest of fuel economy and weight factors, sleepers are constructed of fairly lightweight materials. The outer skin of the sleeper is attached to an aluminum framework that is easy to cut with all the standard extrication tools. The outer skin of the sleeper can be made of aluminum or fiberglass both of which can be removed with standard tools. As a matter of comparison, entry into a sleeper is similar to performing a three-sided roof cut on an automobile on its side. In both cases, a hole large enough to work in is made but not so large that the process takes a long time. When using an air chisel, the outer skin is first removed to expose the structural members that are then severed. When using a reciprocating saw, the outer skin and structural members can be cut simultaneously.

The vehicle's position when it comes to rest will determine which part of the tractor is easiest to enter. If the vehicle is pulling a trailer

Fig. 9–9 Interior view from the driver's seat of sleeper bunk beds.

Fig. 9–10 This sleeper module is equipped with an access door located at the end of the bed.

and rolled over onto its side, the roof may be the best choice. If the vehicle has stayed up on all its wheels, a side entry will provide the easiest access. If the tractor isn't pulling a trailer, entry through the rear wall may be a good choice. When removing portions of the side, roof, or rear of the vehicle, similar type supporting structures are encountered and treated in the same manner.

To simplify the process and reduce the number of cuts needed to enter a sleeper, existing openings, such as windows and vents, should be exploited. These openings can also be used to size up the interior before the first cuts are made. When attempting to enter through a roof or side of the vehicle, a good size-up may reveal that the aerodynamic components of the tractor are actually add-ons that can be easily removed exposing the true outer surface of the roof or sides.

As the size of the sleeper increases, so does the likelihood that interior will contain all the luxuries of home. Many sleeper compartments are so elaborate that the manufacturers use terms like *condo*, *hi-rise*, and *sky rise sleepers*. As the sleepers become more home-like, crews making entry through the walls are more likely to encounter shelves, closets, and bunks that will complicate the entry. For the sake of comparison, forcing entry through an elaborate sleeper could be similar to forcing entry through the walls of a nice travel trailer.

When performing some evolutions, being able to identify the location of the aluminum framework beneath the skin may be useful. Fortunately, the aluminum framework of sleepers built with aluminum skins can be identified by the long row of fasteners. These fasteners can be rivets or huckbolts and are used to attach the outer skin to the framework.

By studying photographs of tractor trailer crashes in the newspaper or on the Internet, it can be seen that, in some situations, the cab and the sleeper will separate during the crash making it easier to gain access to the occupants. Even if these openings can't be used for access and removal of the victim, they can be used for size-up. When is comes time to remove a victim from the sleeper area, a backboard can be slid down a ladder instead of trying to balance the victim overhead.

Fig. 9–11 The interior framework and access door can be seen in this interior view of the same unfinished sleeper module. The oval component in the access door is an air vent.

Fig. 9–12 Once on a backboard, a folding ladder or roof ladder can be used to lower a victim to the ground.

Trailer Underride by Automobile

In addition to crashes that involve the cab and driver, tractor trailers can also be involved in entrapments when automobiles drive under the side of a trailer in a type of crash known as an underride. Underrides can be difficult because, unlike most crashes where the vehicles make contact and separate, the vehicles involved in an underride make contact and often remain connected and in motion until the tractor trailer comes to a stop. When an automobile is involved in an underride, it is often dragged sideways down the street, causing it to become deeply wedged beneath the trailer.

There are two approaches to extricating victims from this type of crash; the first approach involves lifting the trailer to disengage it from the automobile and then pulling the automobile out from beneath the trailer. The second approach involves tunneling through the automobile to reach the victims. Each approach has its drawbacks and advantages. When performing a lifting operation with a wrecker, the officer is taking responsibility for the outcome of a situation that relies solely on the skills, judgment and equipment of others outside the fire department. This is a very serious situation that, if it goes wrong, could result in the crushing death of those entrapped in the automobile.

A better, safer approach involves tunneling through the automobile to reach and remove the occupants. By using the tunneling approach, the officer has knowledge of the personnel and equipment involved in the operation. Additionally, the tunneling method doesn't involve lifting the trailer, which is a procedure that can result in an undesirable outcome should the lifting equipment fail during the lift.

Summary

Most of the steps used in automobile extrication cross over to truck extrication. By recognizing the similarities between trucks and automobiles, crews working on a tractor trailer or truck will be able to intuitively adapt to the problems they encounter. Effective adaptation is the result of sound understanding of extrication principles and techniques. The differences that are encountered will require a good size-up and application of real-world tactics to maintain a safe and effective scene for the victims and crew alike. While the size of the vehicles can be intimidating, a good size-up, as well as planned and effective application of extrication tactics will help bring even a complex truck extrication incident to a successful conclusion.

Procedures

Procedure 1
Opening and Removing Conventional Hoods

To release this type hood latch, the latch is first pulled upward.

Then pivoted outward.

This rubber latch is released by lifting the lever.

After releasing the latches, the handhold on the front of the hood is grasped.

The hood is then pulled to the fully open position.

Springs or other devices prevent the hood from striking the ground.

Trucks and Tractor Trailers

Procedure 1 *(continued)*
Opening and Removing Conventional Hoods

To allow the hood more forward movement, the stops are severed.

This tractor uses a dual cable and spring system to prevent overextension of the hood.

As the second cable is cut, the hood pivots to the ground.

If it's necessary to completely remove the hood, the pivot hardware can be severed.

Hydraulic cutters or an air chisel can be used to cut through the components.

The ability to cut the pivoting parts is based on the cutting force of the cutters.

Vehicle Extrication: A Practical Guide

Procedure 2
Truck Stabilization

When a small amount of stabilization is needed to prevent any rocking motion or additional movement of a tractor, Capabears or other strut devices can be used. It is important to evaluate the load that will be placed on the device and the contact points on the body of the vehicle.

In this example, the head of the Capabear support is placed against the edge of an access door opening. The design and construction of this type of opening may not be able to support a substantial load should one be placed on it. It would be better to consider this type of support useful in limiting a small amount of rocking but not for situations that could result in a substantial load being placed on the support.

If the device is used, the tensioning strap can be attached to a frame member in line with the support.

Trucks and Tractor Trailers

Procedure 2 *(continued)*
Truck Stabilization

A combination of procedures may be needed to reduce the chance of movement if the vehicle is at risk of slipping or moving in some manner.

In this example, the vehicle is tied back with a chain come-along. To create an anchor point for the anchor chain, a pry bar was passed through one of the manufactured holes in the barrier wall.

To assist in preventing rocking, a few well-placed pieces of cribbing and wedges may be used. A 6x6 with a pair of married wedges were able to snug up the front bumper and help eliminate any rocking movement.

Vehicle Extrication: A Practical Guide

Procedure 2 *(continued)*
Truck Stabilization

If the cab has rocked forward and broken loose from the locks that attach it to the chassis, a ratchet strap can be used to tie the cab down to the frame.

A hole can be punched through the side of the vehicle, preferably just above a horizontal structural member, then run down and under the chassis to the other side of the vehicle. In this example, the metal fasteners indicate the presence of a support member.

Another option is to use a chain to attach the cab to the frame rail. The chain can be looped around the selected members and shortened to the appropriate length with the chain shortener. If the chain is a simple extension chain consisting of the chain and a grab hook at each end, one of the grab hooks can be used to shorten the chain to the appropriate length.

Trucks and Tractor Trailers

Procedure 3
Removing a Windshield with Hand Tools

Removing a truck windshield is basically the same as removing an automobile windshield. A windshield saw or ax can be used to cut around the perimeter of the windshield, cutting away the glass and soft laminate.

The top, side, and bottom of the glass are cut on one side then the other.

When removing the glass on a conventional cab design, the hood is first opened to allow the crewmembers to reach the windshield by climbing up on the tire or engine. When the windshield is being removed on a cab-over truck, a ladder will be needed to reach to higher parts of the windshield.

Vehicle Extrication: A Practical Guide

Procedure 3 *(continued)*
Removing a Windshield with Hand Tools

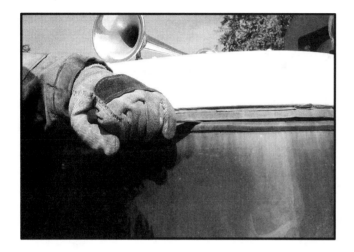

If for some reason the glass is intact and needs to be removed, a sharp knife can be used to cut through the windshield rubber gasket that connects the glass to the window frame. To make the cut, the knife blade is inserted in the small gap between the glass and the windshield frame. The gap won't be visible, but the approximate location can be estimated by splitting the distance between the frame and the glass. If glass or metal is encountered, a slight repositioning will put the knife blade in the right spot. Once all the way though the gasket, the knife is pulled or pushed to cut around the window.

The gasket may consist of two pieces: the gasket and a rubber bead that is pressed into place. The rubber bead is pressed into a groove in the gasket after the windshield is in place, snugging up the installation. To remove this type of windshield, the installation order is reversed. A knife or slot head screwdriver is used to pry some of the bead out of the gasket groove, then the bead is pulled out of the groove all the way around the window.

Once the bead is removed, a small tool may be needed to pop the bead out of the gasket before removing the windshield.

Trucks and Tractor Trailers

Procedure 4
Popping a Door Open with an Air Chisel

After a crash, the height and access to a door may make it difficult to force open with hydraulic spreaders. An alternative is to use a high-performance air chisel to cut the striker and release the door. To cut most strikers, the air chisel should be designed to operate at about 250–275 psi. If the gap is tight between the door and the latch pillar, a Halligan and ax can be used to open up the gap.

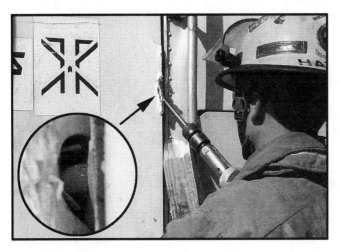

A sharp curved chisel bit is used to cut the latch. The shape of the bit allows it to be inserted in the small gap between the door and the latch pillar. The tool operator will have to be in a good position to apply the required amount of pressure behind the tool needed to cut through the striker.

This striker is similar to those used on automobiles and was cut in a similar amount of time (1–2 minutes). After the striker is cut, the door can be pried open with a Halligan.

Procedure 4 *(continued)*
Popping a Door Open with an Air Chisel

This steel cab and door has a larger gap that may not have to be expanded with a Halligan.

While there are two rods in this striker design, they are smaller than the more common single design.

With the air chisel set at the appropriate pressure, they can be cut just as easily.

Procedure 5
Popping a Door Open with Hydraulic Spreaders

Popping open a door of a cab-over is difficult because of the height of the door and its latch. Unlike the door latch of a standard automobile, this door latch is head high, making it difficult to reach with spreaders.

Instead of attacking the latch area first, a good purchase point can be created in the lower corner of the door. The shape of the door may make it difficult to get a good bite. A wedge can be used to maintain any gap that is initially created with the spreaders.

With the wedge in place, the spreaders can be repositioned for a better bite. If the door is sturdy, the door may stay intact and break free from the latch.

Vehicle Extrication: A Practical Guide

Procedure 5 *(continued)*
Popping a Door Open with Hydraulic Spreaders

If the door fails to pop open cleanly, the wedge can again be used to maintain the opening created by the spreader as it is worked up the gap toward the latch.

In this example, the outer door metal has separated from the inner door structure. The inner door layer has remained attached and inaccessible as the outer door layer is spread outward. At this point, there are two options to opening the door. The first option is to view the latch and striker to see if it is possible to cut it with a high-performance air chisel or the hydraulic cutters. The second option is to position the spreaders above the latch and to spread again. For safety reasons, it is best not to try to operate hydraulic spreaders above shoulder height, making the use of the hydraulic cutters a better choice.

When cutting the striker with a cutter, the striker should be positioned in the jaws as deeply as possible to maximize the cutting force.

Procedure 5 *(continued)*
Popping a Door Open with Hydraulic Spreaders

The latch on this conventional cab is a little lower than the cab-over, making use of the spreaders a little easier. The shape of this steel door allows a purchase point to be created using the pinch and curl method. The spreader tips are opened enough that one tip can be placed on the inside of the door lip and the other tip on the outside. The tips of the spreaders are closed tightly to get a firm grip on the small piece of sheet metal at the corner of the door.

With tips firmly clamped down on the corner of the door, the spreaders are walked toward the front of the vehicle.

The gap created is often large enough to serve as a good purchase point for the spreading procedure, or the pinch and curl can be performed again to create a bigger opening.

Vehicle Extrication: A Practical Guide

Procedure 5 *(continued)*
Popping a Door Open with Hydraulic Spreaders

After starting the spreading operation, you can see the shape and depth of the door. The shape of the area makes it difficult to get a deep purchase point on the door, making continued spreading of the door the best option.

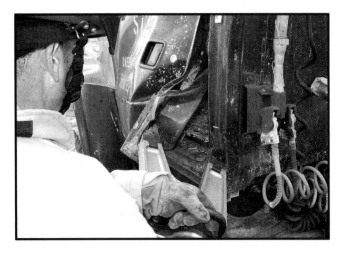

As the spreader opens, a large gap is created that can be used as an improved purchase point. The spreaders are closed, reinserted, and opened again.

As the lower portion of the door is spread away from the jamb, the spreaders can be moved up toward the latch. Once the latch and striker are exposed, the spreaders can continue to spread or cutters can be used to cut the striker.

Trucks and Tractor Trailers

Procedure 6
Removing a Door at the Hinges with an Air Chisel

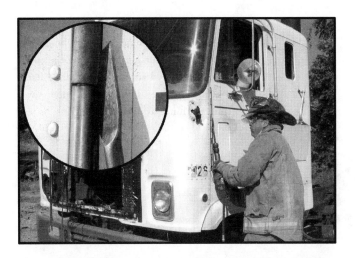

The air chisel is the best choice of tools when it is necessary to remove a door at the hinge when the door is attached with a piano hinge. A sharp, curved cutter blade is used to cut from one end of the hinge to the other.

To pull the flush-mounted door out of the door opening, the adz of a Halligan is driven between the door and the hinge pillar.

In this example, the door check was made of cloth that was cut with a knife, but metal stops may be encountered and can be cut with the air chisel.

Vehicle Extrication: A Practical Guide

Procedure 7
Removing a Door with Hydraulic Tools

Removing a door that utilizes a piano hinge instead of standard automotive style hinges can be done with hydraulic tools but not as easily as when removed with an air chisel. To start, the spreaders are used to tear the door away from the hinge as much as possible.

The lightweight construction of the body may fail before the hinge separates. Additionally, the door stop located on the interior side of the door may create some resistance.

Instead of allowing the metal to fail and tear, the cutters are used to cut the piano hinge instead of trying to spread it apart. The length of the cut can be maximized by opening the cutters completely and positioning the hinge deeply in the cutter opening.

Procedure 7 *(continued)*
Removing a Door with Hydraulic Tools

After cutting as much of the hinge as possible, the spreaders are used again to open up the gap between the hinge pillar and the door for the repositioning of the cutters.

If the gap opened up with the cutters is too small for repositioning of the cutters, the tool operator may have to use the tool from the other side of the door or attack the hinge from the top. Coordination between the cutter and spreader operators will allow each tool to be used to best advantage.

Throughout the procedure, another crewmember can help by positioning the door as it is cut away and prevent it from falling on the tool operators when the hinge is completely severed.

Vehicle Extrication: A Practical Guide

Procedure 7 *(continued)*
Removing a Door with Hydraulic Tools

If the truck door is outfitted with automotive-style hinges, powerful hydraulic cutters should be used when possible to minimize the amount of metal deformity and for overall speed of the operation. The ability to perform this procedure is based on the power of the cutters being used. This type of action will tax a cutter so it is important to ask the dealer's representative about the recommended limitations of the equipment before using it. When using the cutters to cut a hinge, place the hinge deeply in the cutter opening to obtain the maximum cutting force.

A second crewmember is used to maintain control of the door during the procedure. Once the top hinge is cut, the crewmember holding the door can pull downward on the outer edge of the door to open up the gap between the door and the forward hinge pillar. As the door is pulled down, the metal of the door, pillar, and hinge will bend, resulting in more space for the cutter to cut the bottom hinge.

The bottom hinge is then cut and the door removed. The door may be equipped with wires or a door check that may need to be cut away before the door can be lowered to the ground.

Procedure 7 (continued)
Removing a Door with Hydraulic Tools

If spreaders instead of cutters are used to remove a door at the hinges, the spreaders are positioned close to the hinge to break it. When removing a door on an automobile, usually the bottom hinge is broken first; but when performing the procedure on a truck, the sequence of hinges is not important. When cutting the top hinge, the spreader tips are placed on top of the hinge instead of the bottom or middle of the hinge. The top of the hinge is preferred because it is the weakest of the three points.

After popping the top hinge, the door check can be cut away or spread away.

The bottom hinge is then broken away. After breaking the top hinge and the door check, the top and the bottom of the lower hinge will exert an equal amount of resistance to the spreader as it opens. In most cases, placing the spreader on top of the bottom hinge will be the best choice of positions. The top position is preferred because there will be a larger gap available for insertion of the tips and less chance of the tips falling out of position as they are opened. As the door hinge is spread apart, another crewmember can prevent the door from coming down on the tool operator after it is cut.

Procedure 8
Pedal Removal with Pneumatic Tools

When a whizzer is used on aluminum pedal assemblies, the sparks associated with cutting steel assemblies are decreased. Cutting through an aluminum pedal assembly is a slow process requiring a lot of air but can be part of a back up plan when other tools won't free the victim's foot.

A more powerful choice of pneumatic tools is an air chisel fitted with a sharp, curved cutter bit. After sizing up the construction of the pedal assembly, you can cut away the weakest or smallest point.

In this case, the aluminum base provided the best access and was easily and quickly severed. Once one side of the base was severed, the pedal assembly was easily taken apart.

Trucks and Tractor Trailers

Procedure 9
Pedal Removal with Hydraulic Tools

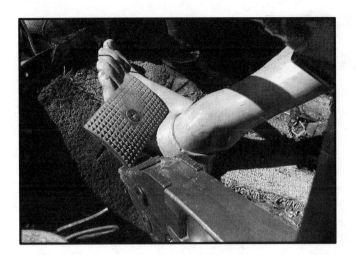

The brake pedal design on this cab-over highway tractor consists of a pedal that is attached to an arm that exits through the floorboard behind the pedal. If the pedal is jammed in the down position, trapping the victim's foot beneath the pedal, it may be possible to insert the spreader tips between the floor and the rear of the pedal and spread the pedal upward to free the foot.

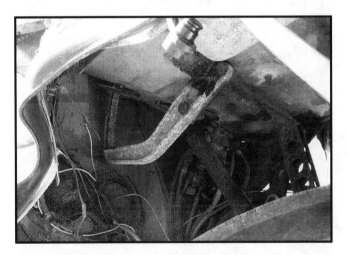

Another option is to view the pedal assembly beneath the cab and identify the location of the problem.

If the assembly is bent or otherwise jammed, hydraulic cutters can be used to cut the assembly away to release the pedal.

Procedure 10
Cutting Away a Steering Ring with Hydraulic Cutters

If a steering wheel is obstructing access or removal of a victim, removing the steering wheel may provide enough space to remove the victim. A ladder may be needed to access the wheel in a cab-over, but flipping the hood open or working off the hood may be possible on a conventional cab. To make the cuts, the cutter is positioned as close to the hub as possible. The spoke is positioned deep in the cutter jaws for maximum cutting power.

To make the subsequent cuts, reposition the cutters on the other spokes. It may become necessary to force the steering wheel to turn to make the final cut.

With the large truck steering wheel out of the way, access to the driver compartment is improved. Depending on the available access, another option is to cut the outer ring of the steering wheel instead of the spokes.

Trucks and Tractor Trailers

Procedure 11
Cab-over Dash Roll with Pneumatic and Hand Tools Only

A dash roll can be performed with two of the most basic extrication tools: an air chisel and a Hi-Lift jack. To sever the roof from the "A" post, a curved chisel bit is used in a position high on the post. When cutting the post with the curved cutter, it is important to cut around the perimeter of the post before trying to cut the metal in the center. If the tool operator attempts to cut straight through the center of the post, the chisel bit will become jammed, complicating the procedure.

With the door cut away, a relief cut can be easily made in the forward hinge pillar. Again, cutting around the perimeter of the structure will prevent the bit from becoming stuck in the metal. To reduce the chance of the bit becoming stuck, only the inner half of the cutting edge should be used to make the cut.

This bit was allowed to become buried in the kerf as the cut was being made. When a bit becomes jammed like the example, it is best to disconnect the air chisel from the bit, insert a new bit, then cut the original bit out of the metal. If it appears to be hopelessly jammed, leave it in place continue the cut around it. The jammed bit can be recovered after the dash is rolled, and the relief cut has opened up.

Vehicle Extrication: A Practical Guide

Procedure 11 *(continued)*
Cab-over Dash Roll with Pneumatic and Hand Tools Only

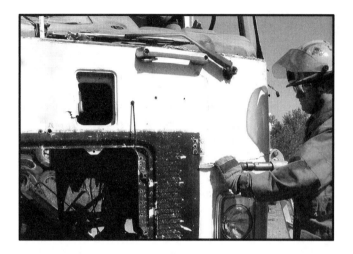

As the relief cut is completed in the forward hinge pillar, the cut is continued to further weaken the structure. The cut should continue along the floor of the cab toward the center of the cab. This cut will serve as a hinge point when the jack is put into action.

The main supporting structure of the dashboard is cut next. The supporting beam runs from one door to the other and provides a lot of the strength in the dashboard. The vertical cut initially cuts the outer layer of sheet metal, then as much of the interior material as possible.

While working off a ladder, the tool operator continues to cut away at the primary structural member of the dash. While each truck model may be different, the tool operator should be able to size up the dash and decide which pieces of metal are providing strength to the dash and which are not. Working in close proximity to the victim will require a good amount of skill on the part of the tool operator to avoid making contact with the victim.

Trucks and Tractor Trailers

Procedure 11 *(continued)*
Cab-over Dash Roll with Pneumatic and Hand Tools Only

With the structure of the cab weakened, the Hi-Lift jack is positioned low on the forward hinge pillar. Low positioning is necessary because of the limited useful length of the jack and the need for the operator to reach and operate the handle. The nose (lifting part) of the jack is positioned in a sturdy area that won't slip or tear as the jack is operated.

In this case, the base of the jack was placed against the structure beneath the seat. This position allows more of the jack to be used for pushing than if the base were placed at the lower rear corner of the door opening.

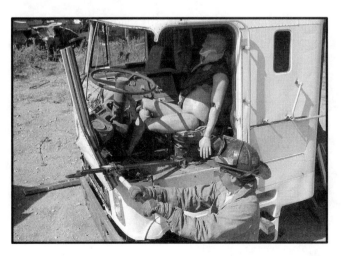

As the jack is operated, the tool operator continually monitors the nose of the jack and stops before it slips out of position. If the jack operator starts to run into resistance while rolling the dash, the air chisel operator can locate the cause of the resistance and sever it before it is put under heavy load.

Procedure 12
Tenting a Roof with Hydraulic Spreaders

When a roll-over crash has left the roof flattened down to the dashboard, quick access to the occupant compartment can be gained by tenting the roof up. This tractor is situated on its driver's side, with the greatest amount of roof flattening in the driver's area. To tent the roof, the spreaders are placed between the top edge of the roof and the dashboard, then opened.

After opening to the maximum spreading distance of the spreaders, the tool is closed, repositioned, and opened again to the maximum opening distance of the spreaders.

The process is continued until the space between the dashboard and the windshield is opened as far as necessary.

Procedure 13
Roof Entry with a Reciprocating Saw

When viewed from the front, the components that make a tractor aerodynamic may seem massive, making a roof entry impossible.

A good walk-around and size-up may reveal that the components are actually add-ons to a standard highway tractor, making it possible to remove the roof fairing to gain access to the top of the cab. In this photo, the crewmember in the fairing has chosen to cutting the attachment points from inside the fairing.

Methods and types of attachments may vary, in this case, the airfoil is attached with small pieces of steel and bolted in place.

Vehicle Extrication: A Practical Guide

Procedure 13 *(continued)*
Roof Entry with a Reciprocating Saw

After the mounting hardware is cut, the airfoil can be removed, revealing a standard cab roof.

If access to the interior can not be made by a crewmember, an inspection hole can be made before the larger roof cuts are made. The inspection hole should be done at a level high enough to reduce the chance of hitting a victim that may be lying at the bottom of the tractor. In this case, an air chisel was used to remove a small piece of sheet metal, but a reciprocating saw can be used just as easily by first creating a starter hole with a Halligan and ax.

After removing the outer sheet metal of the inspection hole, the ceiling material and insulation can be cut away.

Trucks and Tractor Trailers

Procedure 13 *(continued)*
Roof Entry with a Reciprocating Saw

It is best to start the cuts with a long reciprocating saw blade so all structural members that are encountered can be completely cut.

Instead of making the four cuts needed for a complete removal of the sheet metal, three cuts can be made, leaving a smooth bend at the bottom of the flap. The smooth edge poses less danger to crewmembers and the victim during the extrication. To start the crease in the sheet metal, a sledge hammer or ax may be needed to strike the area.

One disadvantage of the three-sided roof flap becomes evident when the outermost edge of the roof flap does not reach the ground. The crewmembers can step on it to bend it down, or a fourth cut can be made to totally remove it from the cab.

Vehicle Extrication: A Practical Guide

Procedure 14
Roof Entry with an Air Chisel

The large flat surface of a tractor roof makes an air chisel a good choice of tools when entry is to be made by cutting and removing a large section of metal. Similar to the roof procedure used for automobiles, make three cuts and flap down the roof skin, or a fourth cut can be made and the flap removed. The sequence of cuts is unimportant, but it is best not to leave the bottom cut for last to avoid having the metal fall on the tool operator's head. In this case, the top cut was done first.

The second cut is made to maximize the size of the finished hole while avoiding reinforced structural members. Based on need and the construction of the roof, the cut can be made in front of or behind the clearance lights.

In this example, to avoid cutting through additional reinforcement or structural members, the downward vertical cut stopped before a long row of metal fasteners that indicate the presence of framework or overlapping metal. The location of the fasteners has been enhanced in the photograph for clarity.

Procedure 14 *(continued)*
Roof Entry with an Air Chisel

As the fourth cut is completed, the sheet metal skin will remain in place because it is attached to the stringers and ribs of the roof that are still intact.

Instead of trying to separate the roof skin from the supporting members, the members are exposed and severed. To expose the supporting member, the sheet metal can be folded over with the fork of a Halligan.

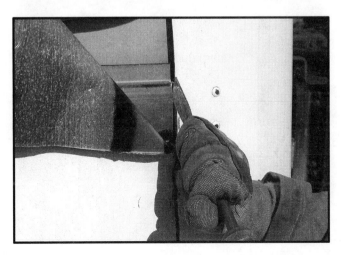

The curved chisel is used to cut through the supporting member.

Vehicle Extrication: A Practical Guide

Procedure 14 *(continued)*
Roof Entry with an Air Chisel

When structural members in the center of the roof need to be cut, there won't be a corner that can be folded down to expose the member, making removal of a piece of metal necessary. Four cuts are made to expose the stringer, with one of the four cuts being the long vertical cut. Three additional cuts are made to expose the support.

After the fourth cut is completed, it may be necessary to cut beneath the sheet metal to sever the fastener.

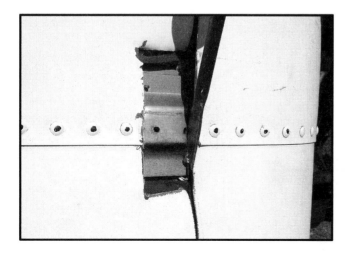

With the sheet metal removed, the support can be visualized and cut with the curved chisel bit.

Trucks and Tractor Trailers

Procedure 14 *(continued)*
Roof Entry with an Air Chisel

If possible, a crewmember is sent inside the cab to size up the interior. By sizing up the interior, you can exploit weaknesses and clear areas and avoid congested areas. Entry through the windshield provides quick and easy access to the interior of the cab.

Once inside, while medical care is being provided to the victim, the inside crewmember can start removing decorative trim from the roof. A variety of materials may be used on the ceiling of the cab, but they all will be lightweight in nature. A razor knife of similar tool may be needed to remove them.

As the ceiling material is removed, the work of the outside crewmember can be seen. By removing the inside and outside layers of the roof, the supporting structures are more easily seen and removed.

Vehicle Extrication: A Practical Guide

Procedure 14 *(continued)*
Roof Entry with an Air Chisel

For easier cutting of structural members, the air chisel can be passed to the interior crewmember. If conscious, the victim should be advised of the increased noise level caused by the operation of the air chisel.

With the ribs and stringers severed, the roof can be removed; or if only the top and two vertical cuts were made, it can be folded down.

If four cuts were made for a total removal, crewmembers should take care not to injure themselves or the victim on the sharp edge of the lower cut.

Procedure 15
Dash Roll-up with a Reciprocating Saw and Come-along

The basic principles of a dash roll on a cab-over tractor can be applied when using a variety of tools as long as a few details are properly addressed. One of the steps for prepping the vehicle for the dash roll involves separating the roof from the forward hinge pillar. When using a reciprocating saw, the cut is made high on the "A" post and at an angle that will allow the post to separate without binding.

This cut was made so the post can move forward without binding. If the cut were made perpendicular to the post or angled in the opposite direction, the post would bind and resist the forward movement of the dashboard.

A relief cut is made at the floor board level with the door still attached at the hinges. The saw is positioned to cut from the interior toward the exterior of the forward hinge pillar. After completing as much of the cut as possible, the tool operator repositions as necessary to extend the cut deeper into the hinge pillar.

Vehicle Extrication: A Practical Guide

Procedure 15 *(continued)*
Dash Roll-up with a Reciprocating Saw and Come-along

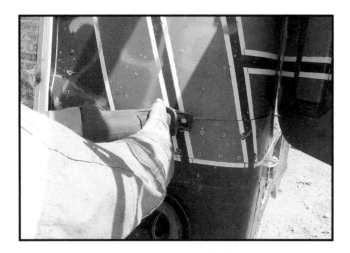

The cut is extended across the front of the cab as far as possible to weaken the structure as much as possible. To avoid making contact with the victim's feet, the tool operator should have another crewmember monitor the cut from the inside of the vehicle.

A long reciprocating saw blade is inserted in the saw. A long blade is one about 12 in. long. The windshield is removed or a hole is made for the positioning of the saw on top of the dash. Once in position, the cut is made straight down through the dashboard in an attempt to cut the main supporting structural member that runs across the dashboard from one door to the other.

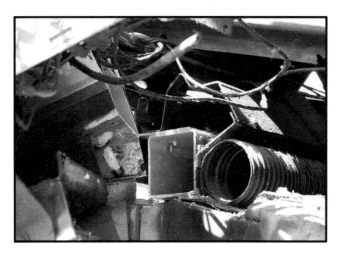

This interior view shows the structural members that have been cut by the saw. If these members are not cut on a cab-over tractor, resistance will be encountered.

Procedure 15 *(continued)*
Dash Roll-up with a Reciprocating Saw and Come-along

A typical dash roll-up involves using a spreading tool, usually a hydraulic ram, to push the dashboard away from the occupant compartment. An alternative to pushing the dashboard away is to pull it away. If the door is intact, it can be used as the attachment point for a chain, and a come-along can be used to pull the door away. The first step in attaching the chain to the door is to pass the sling hook through the window opening.

The hook is positioned half way down the interior of the door, with the open part of the hook facing the door hinge.

The chain is placed in the hook, creating a loop around the door. The loop should be positioned over the inner third of the door.

Procedure 15 *(continued)*
Dash Roll-up with a Reciprocating Saw and Come-along

A little tension is applied to the chain to keep it in place while the chain is brought around the outer edge of the door.

The chain shortener is attached at the outer edge of the door. The completed wrap should be identical to the door widening procedure used in the automobile section.

To pull the dashboard away from the victim, a 2-ton come-along can be used. The come-along is attached to the master link of the door chain and walked back to the anchor chain. The anchor can be located directly in front of the tractor or toward the opposite side. Pulling straight ahead would ensure best results, but, in this example, the anchor will be established on the opposite side of the vehicle.

Procedure 15 *(continued)*
Dash Roll-up with a Reciprocating Saw and Come-along

Simulating a situation where there is no anchor point available at the front of the tractor, an anchor chain was hooked up to the chassis on the passenger side of the vehicle. A crewmember has shortened the anchor chain so the master link is located at the front corner of the cab.

The come-along is attached to the master link and cribbing put in place to distribute the load that will be applied by the come-along as it is put under tension. With rigging completed, the officer directs the come-along operator to take the slack out of the system. With the slack out, the crewmembers check their rigging for any problems that may have occurred as the slack was taken out.

With the system checked, the come-along is operated. The first action will be the movement of the door to its full open position. After the door is fully extended, it will move beyond full extension. As the door moves, the hinges should be monitored for possible failure.

Vehicle Extrication: A Practical Guide

Procedure 15 *(continued)*
Dash Roll-up with a Reciprocating Saw and Come-along

The anchor chain is monitored and additional cribbing applied as necessary.

After the door has overextended its hinges, either the hinge will fail or the sheet metal of the front of the cab will fail. If an adequate relief cut was made at the beginning of the operation, the sheet metal of the cab should fail before the hinge.

As the door is pulled, the front of the cab and the dashboard will follow it in the direction of the come-along cables.

Procedure 16
Dash Roll-up with Hydraulic Tools on a Cab-forward

The basic steps of rolling a dash with hydraulic tools on a cab-forward tractor are similar to those used on automobiles. For the dash to move forward, it must be disconnected from the roof. Flapping the roof is impractical on a truck because of the height, but severing the "A" post is a good alternative. By cutting high, there is less of the post obstructing the passenger compartment after the dash is rolled forward.

The structural integrity of the forward hinge pillar is further weakened by cutting into the pillar as deeply as possible. In this example, the door has been removed. If the door is left on the truck, the relief cut should be made beneath the bottom hinge. If the cut were made between the hinges of an intact door, the door would act as a splint and prevent the relief cut from opening up.

To maximize the amount of space created, the relief cut is made at floor level.

Procedure 16 *(continued)*
Dash Roll-up with Hydraulic Tools on a Cab-forward

With the cut completed, and the design strength derived from the shape and multiple layers of metal eliminated, only the relatively flat metal of the front end remains.

The largest ram that will fit in the door opening is selected and placed in the door opening to check for a proper fit. The base of the ram is positioned in the corner of the door and the head of the ram is positioned at an angle that will prevent it from sliding up and off the forward pillar as it is extended. If the ram fits in the space in the retracted position, it is hooked up to the hose.

As the ram is extended, the tool operator should constantly evaluate the contact point between the head of the ram and the pillar and stop extending the ram before the angle causes the head to slip off the pillar.

Procedure 17
Dash Roll-up with Hydraulic Tools on a Cab-over

This cab-over has more size and structure than similar-looking trucks used for local and around town deliveries, but many of the steps are the same.

To disconnect the roof from the front of the tractor, the "A" post is cut.

As is done with the smaller trucks, the forward hinge pillar is cut to remove the strength and structural integrity from the front of the vehicle.

Vehicle Extrication: A Practical Guide

Procedure 17 *(continued)*
Dash Roll-up with Hydraulic Tools on a Cab-over

The cut is made as deeply as possible to remove as much of the curved and reinforced metal as possible.

The cut is made as close to the floor as possible to maximize the amount of metal that will be pushed out of the way as the ram is extended. When setting up the cut, the tool operator can simply lay the inner blade directly on the floor. The reinforced areas that are being weakened can be identified by the fasteners.

An important part of the dash roll procedure is weakening the dashboard assembly. Hydraulic cutters are used to sever the primary dashboard support that runs from one door to the other. By repositioning the cutter as needed, the support can be completely severed.

Procedure 18
Dash Roll-up with Hydraulic Tools on a Conventional

The ram head is positioned on the hinge pillar in an area that will allow it to get a good bite and not slip off as the ram is extended. Rubber weather stripping can cause the ram head to slip and can be removed before the ram is put in position.

The base of the ram is positioned at the bottom corner of the door. If given a choice of positions, the base of the ram should match the contour of the vehicle to maximize the contact area between the ram and the vehicle. As the amount of contact between the ram and the vehicle increases, the likelihood of the metal tearing is decreased, providing a better push-off point.

As the dash is rolled forward, the sheet metal severed by the relief cut may continue to tear or the body members may simply pull apart.

Vehicle Extrication: A Practical Guide

Procedure 18 *(continued)*
Dash Roll-up with Hydraulic Tools on a Conventional

The procedures for rolling the dash on this conventional cab (engine in front) are similar to the other trucks. To simplify the dash roll, the hood is tilted out of the way.

The "A" post is cut high to separate the roof from the firewall and dashboard area.

Understanding the principles of rolling a dash is helpful when trying to identify any cuts that may need to be made. The key factor in the success of any dash roll-up is the proper weakening of the structures that support the dashboard and hold it in place. By applying that concept to this vehicle, the support assembly running from the firewall to the radiator is easily identified as the component that must be defeated for the dash to roll forward.

Procedure 18 *(continued)*
Dash Roll-up with Hydraulic Tools on a Conventional

Initially it would seem that severing the support would allow the dashboard to roll forward. However, as the dash moves up and forward, the cut ends of the support would meet and create resistance. One solution would be to make a second cut at the radiator, removing the entire support.

Another option is to make the cuts adjacent to the firewall.

With both the top and the bottom of the support bracket cut, the bracket can be removed, creating a clear area.

Vehicle Extrication: A Practical Guide

Procedure 18 *(continued)*
Dash Roll-up with Hydraulic Tools on a Conventional

A relief cut is made in the forward hinge pillar to separate it from the floor and the rocker panel.

Even with large cutters, the initial cut may not sever all the components in the area.

Additional cuts may be required to cut through the fender and front pillar. This is time well spent since a little extra effort made in weakening the structure before it is rolled can result in a more successful roll-up in any type of vehicle.

Procedure 18 *(continued)*
Dash Roll-up with Hydraulic Tools on a Conventional

This photograph shows the front side of the dashboard and the driver's foot area. The shaft in the left side of the photograph is part of the steering assembly. While it looks sturdy, the steering column is actually connected to a universal joint inside the cab that allows it to pivot, creating little resistance to a dash roll.

Of more significance is the supporting member that extends from the bottom of the cab to the firewall. Regardless of manufacturer and design, a good examination will identify how strength is built into the vehicle, which is the basis for determining where and how the strengthening structures vehicle can be weakened.

To get a good wrap around the supporting member, one of the cutter blades were positioned in the steering assembly hole. To avoid cutting the occupant, the tool operator should have another crewmember view the positioning from the inside of the cab. If unable to see the victim's feet, the cutter should be repositioned and used only in an area where the tool operator can see what is being cut.

Procedure 18 *(continued)*
Dash Roll-up with Hydraulic Tools on a Conventional

The ram is positioned at an angle that prevents it from sliding up the forward hinge pillar as the ram is extended.

The head of the ram is positioned over sturdy metal areas of the hinge pillar.

The ram base is positioned in the lower rear corner of the door opening. The weather stripping can be removed to reduce slippage.

Procedure 18 *(continued)*
Dash Roll-up with Hydraulic Tools on a Conventional

A two-sided dash roll-up was performed on this vehicle with a ram on each side of the vehicle. The relief cut—the weak point in the structure—continued to tear as the dashboard rolled up and forward.

The need for removal of a large portion of the front support bracket assembly can be seen here. The firewall has moved through the space previously occupied by the bracket and nearly made contact with the main support. If the bracket or the support had not been removed, the firewall would have made contact with it, creating resistance. This resistance would have caused some other part of the structure to fail, possibly resulting in complete failure of the operation.

The steering assembly pivot point provided no resistance to the roll-up. As with the dash roll-up procedure used on automobiles, one of the good effects of the roll-up is the upward movement of the pedals that can result in disentanglement of the victim's feet.

Vehicle Extrication: A Practical Guide

Procedure 19
Dash Lift with Hydraulic Tools on a Cab-over

In situations when a dash roll-up is not possible because of obstructions in front of the cab that restrict access or movement of the cab, a dash lift may be performed. Unlike the dash roll-up, the dashboard will move upward and away from the victim. This upward movement will necessitate the complete removal of the "A" post instead of simply severing it as is done with the dash roll procedure. The first cut is made high on the "A" post.

The second cut is made low on the "A" post.

The severed portion of the "A" post is then removed.

Procedure 19 *(continued)*
Dash Lift with Hydraulic Tools on a Cab-over

To create an area for the spreader tips, make two horizontal cuts in the forward hinge pillar, a few inches apart.

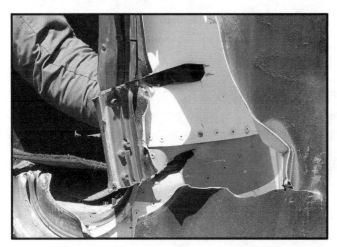

One cut is made at the floor board level and the next cut is made above it.

To create a good purchase point for the spreader tips, the metal between the two horizontal cuts needs to be moved out of the way. The easiest way to move the metal out of the way is to first open the spreader tips a few inches, then close them, clamping the metal between them.

Vehicle Extrication: A Practical Guide

Procedure 19 *(continued)*
Dash Lift with Hydraulic Tools on a Cab-over

The spreaders are then walked toward the front of the vehicle, causing the metal to bend at the end of the cuts.

The spreader tips are then repositioned vertically in the gap.

The spreaders are opened, forcing the dash and the front of the cab upward and away from the victim.

Procedure 20
Cab-over Dash Removal with a Reciprocating Saw

The passenger side of a cab-over tractor has less wiring and components than the driver's side, making total removal of the dash an option. The procedure is similar to the dash roll with the difference being that the cuts are complete instead of forming relief cuts. The "A" post is cut high at an angle that will allow the post to move forward without binding.

The forward hinge pillar is cut at floor level with a 10–14 TPI reciprocating saw blade.

The size of the forward hinge pillar structure can be seen in comparison to a reciprocating saw blade. A short blade can be used for the floor level cuts, but a longer blade will be needed for the dashboard cuts.

Procedure 20 *(continued)*
Cab-over Dash Removal with a Reciprocating Saw

To avoid having the blade make contact with the victim's feet or get bogged down in the floor board, a crewmember can assist the tool operator by observing and directing while viewing the inside of the cab. The cut extends to the center portion of the cab.

With a 12-in. blade on the saw, the vertical cut is made straight down through the dashboard.

While working off a ladder, the tool operator will have to make the cuts from whatever area gives the best access to the dashboard and the main structural supporting member in the dash.

Procedure 20 *(continued)*
Cab-over Dash Removal with a Reciprocating Saw

Unable to see the position of the blade, the tool operator may again need assistance from a crewmember who can see the interior as the cut is made.

This photo shows the main dashboard structural member after the procedure was finished. The size and proportion of the structure can be compared to the saw blade and the operator's hand.

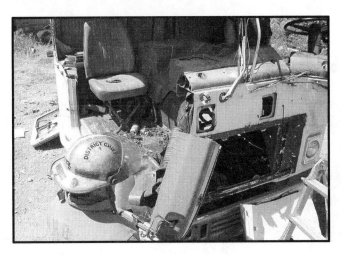

By successfully completing the three standard cuts, you can remove the entire passenger side of the cab and dash, providing unobstructed access to the passenger.

Procedure 21
Sleeper Side Entry with a Reciprocating Saw

Side entry into a sleeper with a reciprocating saw is similar to the air chisel procedure but is simplified if a long blade is used. By using a long blade, both the sheet metal and the structural supporting members can be cut simultaneously. This sleeper has a vent hole that can be used to start the cuts instead of having to punch a starter hole with the pike of a Halligan. The first cut extends from the existing hole, forward to the desired ending point.

The next cut starts again from the vent hole, extending rearward.

The saw is turned at the end of the horizontal cut and transitions into the vertical cut.

Procedure 21 *(continued)*
Sleeper Side Entry with a Reciprocating Saw

As the downward cut is completed, the saw is turned to make the lower horizontal cut.

The cut is extended to the rear of the door frame or can extend completely through the door frame.

In this example, the cut was extended into the door opening, eliminating the need for another vertical cut to connect the two horizontal cuts.

Procedure 22
Sleeper Side Entry with an Air Chisel

Entry into a sleeper from the side with an air chisel may be easier to accomplish than rear entry if the tractor is pulling a trailer. To totally remove the side of the sleeper, four basic cuts will be made, two horizontal and two vertical. The top cut will require the use of a ladder or the bed of a flat bed wrecker positioned to serve as an elevated platform. A curved cutter is the best choice for cutting the sheet metal because of the presence of underlying support members that will slow a T cutter.

As the end of the first horizontal cut is made, the air chisel is manipulated into a curve that transitions into the vertical cut.

The next transition is made into the horizontal cut, ending where the first cut began.

Trucks and Tractor Trailers

Procedure 22 *(continued)*
Sleeper Side Entry with an Air Chisel

Metal fasteners or rivets will help identify the underlying framework or skeleton of the cab. It is possible to cut through these structural members without completely exposing them, but it increases the chances of an incomplete cut or the chisel bit getting hung up in the metal. The best approach is to remove the outer sheet metal to expose the member then cut them. The location of the fasteners in this photograph is emphasized for clarity.

By removing a small amount of sheet metal, you can see the structural supporting members, and it can be accessed and cut with the curved chisel bit.

The process is repeated wherever a structural member is encountered that prevents the removal of the outer layer of sheet metal. After cutting away the supports, the outer sheet metal can be removed.

Vehicle Extrication: A Practical Guide

Procedure 22 *(continued)*
Sleeper Side Entry with an Air Chisel

Any duct work or other lightweight structures found to be obstructing access to the sleeper is cut away with the curved cutter bit.

The interior view of the procedure as it is carried out shows the ductwork vent and the structural support members that are located behind the outer sheet metal.

When planning the size of the hole in the sleeper, consider the size of a long backboard.

Procedure 23
Rear Entry into a Conventional Steel Cab with a Reciprocating Saw

The rear wall of this type of conventional steel cab can be removed with a reciprocating saw. If a window were not present, four cuts would be required to remove the section of double layered steel. The glass in the window can be removed and the opening can serve as one of the four cuts required to remove the rear wall. To simplify starting the lower horizontal cut, a starter hole can be punched at the beginning of the procedure with a Halligan or, in this example, a windshield saw.

The first cut extends downward from the window opening to the starter hole that was made in the lower part of the wall.

After completing the downward cut, the saw is removed and repositioned in the starter hole for the low horizontal cut. That cut extends to the area of the next planned vertical cut.

Vehicle Extrication: A Practical Guide

Procedure 23 *(continued)*
Rear Entry into a Conventional Steel Cab with a Reciprocating Saw

The second downward vertical cut is started and extends to the ending point of the horizontal cut.

With the cut completed, the wall section can be removed.

While cutting through the wall, a long blade may be a good choice in areas where there are two layers of steel separated by insulation and other components.

Trucks and Tractor Trailers

Procedure 24
Rear Sleeper Entry with a Reciprocating Saw

To remove the sheet metal at the rear of a sleeper with a reciprocating saw, some type of starter hole is needed. The simplest, quickest method is to punch a hole through the metal with the pike of a Halligan. This can be done in any corner of the area to be cut away, but the interior area should be sized up first to make sure the victim is not in the area where the hole will be made. Once the hole is made, the reciprocating saw is positioned in the hole and the first cut started.

To reduce the chance of binding, the bottom cut can be completed as the first or second cut. After the second cut, there is a greater chance of the metal settling downward into the kerf, causing the blade to jam.

For quickest results and to eliminate the need for another starter hole, the first cut can curve and transition into the second cut.

Procedure 24 *(continued)*
Rear Sleeper Entry with a Reciprocating Saw

The second cut can transition into the third cut, again eliminating the need for another starter hole.

As the cuts are made, it is best if a crewmember can be assigned to the interior of the sleeper to assess the victim and remove interior trim. The interior crewmember can coordinate the cuts with the tool operator, identify structural members as they are encountered, and help avoid problem areas.

As the fourth cut is completed, the rear sleeper panel can be removed.

Procedure 25
Rear Sleeper Entry with Hydraulic Spreaders

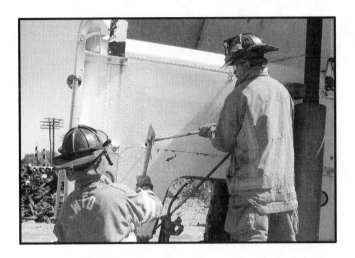

Using hydraulic spreaders to remove the sheet metal of a sleeper may not come to mind as a first choice of tools, but they can be used to open up a hole of limited dimensions. The pike of a Halligan along with a sledge or flat head ax are used to punch a hole through the metal at an inside corner where horizontal and vertical structures meet.

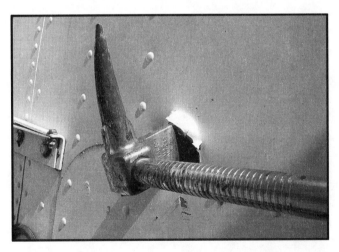

The initial hole is punched with the pike of the Halligan, followed by an enlargement of the hole with the adz. The finished hole should be large enough to allow the insertion of both spreader tips.

In this example, the vertical cut was made first. The spreader tips are placed in the starter hole then opened. The lower tip will spread through the sheet metal until it makes contact with the horizontal structural member. Encountering greater resistance, the lower tip will stop moving after making contact with the member, and the upper tip will start upward tearing through the sheet metal.

Vehicle Extrication: A Practical Guide

Procedure 25 *(continued)*
Rear Sleeper Entry with Hydraulic Spreaders

The spreader is opened to its maximum opening distance.

As the spreader moves upward, the metal will tear and fold, similar to a T cutter on an air chisel.

The second cut is made by closing the arms of the spreader and repositioning them to the original starter hole position. The second cut is a horizontal cut that uses the vertical structural member as a push-off point. Like the vertical cut, the spreader tip that is positioned against the vertical structural member will remain stationary while the other tip tears through the metal.

Trucks and Tractor Trailers

Procedure 25 *(continued)*
Rear Sleeper Entry with Hydraulic Spreaders

The third cut is made by again using the horizontal member as the push-off point, allowing the upper tip to tear through the metal until the spreader is opened completely.

With the interior trim and insulation exposed, it can be cut or pulled out of the way.

While the hole is not very large, it is large enough to gain access to the victim and possibly large enough for a backboard.

Procedure 26
Rear Sleeper Entry with Hydraulic Cutters

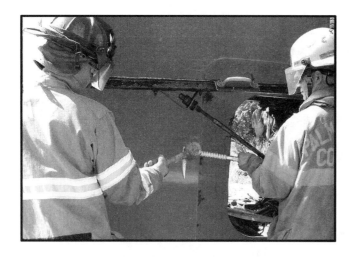

Hydraulic cutters are not the best tools for making long cuts through sheet metal. They are used primarily for severing metal that the cutter blades can wrap around. However, they can be used for cutting sheet metal if the proper technique is used. To cut through the rear of this metal sleeper that is resting on its side, a starter hole is made with a Halligan and sledge or ax. The pike can be used initially, but then the adz is used to create a long hole.

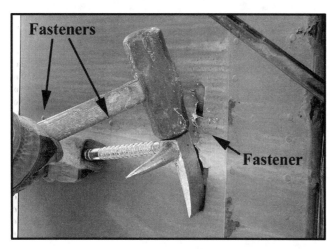

The long cuts are made on each side of a structural member that can be identified by the presence of metal fasteners. The purpose of the long starter holes is to create holes that are long enough for the insertion of the cutter blades. As the back of the Halligan adz is struck, the adz tears through the sheet metal like an oversized air chisel bit.

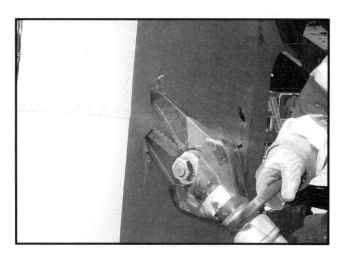

The cutter blades are inserted in the holes and the underlying structural member is cut.

Procedure 26 *(continued)*
Rear Sleeper Entry with Hydraulic Cutters

This interior view of the previous exterior photograph shows the aluminum framework member as it is being cut.

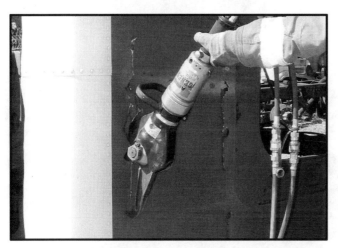

To make the long sheet metal cuts, the cutter is used like an old-fashioned manual can opener. One blade is slid deeply into the hole until the sheet metal reaches all the way to the notch on the cutter. The body of the cutter is pivoted to bring it as close to the metal as possible. As the cutter is closed, the tips of the blades will pierce the metal, starting the cut. The tips will help hold the blades in position while the serrated portion of the blades completes the cut. The cutter is opened and slid into position for the next long cut.

Another option is to use the pike of the Halligan to punch a series of holes large enough to receive the cutter tips. A tip is placed in each hole and the cutters are closed.

Procedure 26 *(continued)*
Rear Sleeper Entry with Hydraulic Cutters

The horizontal cuts are continued until the size of the hole is sufficient or when a large structural member is encountered that is better left uncut.

The third cut is made in the same manner as the previous two cuts.

With three sides of the hole cut, the sheet metal can be flapped down to access the sleeper. If necessary, a fourth cut can be made, completely removing the metal. If cut away with hydraulic cutters, the lower cut will result in sharp, jagged edges.

Trucks and Tractor Trailers

Procedure 27
Trailer Underride

When a car drives under a trailer, both vehicles may be moving, causing the automobile to become securely wedged beneath the trailer when they come to rest. At rest, the vehicle will be taking up a certain amount of vehicle height that will remain constant until extrication procedures are started that can result in a change in the shape, strength, and position relationship between the vehicles. To retain the space between the ground and the trailer, cribbing can be used to box crib the trailer. This action will contribute to maintaining the at-rest height of the trailer.

Unfortunately, most fire apparatus don't carry enough cribbing to build even a single box crib between the ground and the underside of a trailer. Another solution that requires only a couple pieces of cribbing is to place the cribbing between the underside of the trailer and the bottom of the window opening. Yet another option would be to build a box crib between the hood of the automobile and the underside of the trailer.

Another quick way to maintain the trailer height and possibly raise it a few inches is to lower the landing gear. The term *landing gear* is commonly used in the trucking industry to describe the manually-operated jack system that's built into all trailers. As the tractor-trailer is being driven, the landing gear is raised to the retracted position for adequate road clearance.

Procedure 27 *(continued)*
Trailer Underride

The handle used to operate the landing gear is permanently attached to the trailer and is folded away when not being used.

To operate the handle, it is lifted out of its storage bracket.

Landing gear mechanisms are built with two operating speeds that are selected by pushing the shaft in or by pulling it out. On this model, the high speed position is the "in" position; other types of landing gear may be the opposite. The high speed can be used when there's no load on the landing gear, such as when the gear is being initially extended from the retacted position to the ground.

Trucks and Tractor Trailers

Procedure 27 *(continued)*
Trailer Underride

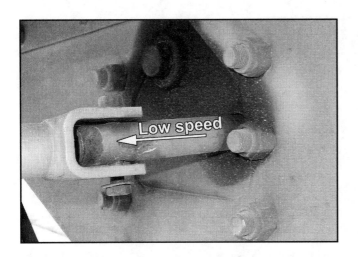

Once the landing gear is put under load, the low speed position can be used for easier operation.

The legs of the landing gear are connected by a shaft and operate simultaneously as the handle is being operated.

Once the relationship between the ground, entrapment vehicle and trailer is stabilized, efforts can be turned to the extrication of the victim. Initial access to the victim can be achieved by a crewmember sliding in through a window, but to extricate and remove the victim on a backgoard, more space will be needed. To create more space, a clear area is created between the front seat and the rear of the vehicle. The first step is to pop the trunk open.

Procedure 27 *(continued)*
Trailer Underride

The trunk lid is removed by cutting the spring-loaded supporting arms. If gas lifters are used to support the trunk, there are several ways of removing them. They can be pried away if connected by a ball and socket connection, the mounting hardware can be cut, or the silver piston can be severed. The pressurized black cylinder should not be cut if it can be avoided.

The metal structure of the rear deck is exposed by removing the trim that covers it.

The structure of underside of rear deck and the rear seat can be seen in this photograph along with the trunk lid spring bars. When sizing up this area, the crewmembers should identify how the entire structure can be cut and removed in the least amount of time and with the least amount of effort.

Trucks and Tractor Trailers

Procedure 27 *(continued)*
Trailer Underride

The trunk lid lifter springs are pulled away, or if necessary, cut away with hydraulic cutters.

The rear deck is the next component that is removed. To minimize the amount of time needed for its removal, a combination of tools can be used based on the type of cut that needs to be made. To cut through the reinforced edge of the rear deck quickly, a hydraulic cutter can be used. A reciprocating saw or air chisel can be used, but a hydraulic cutter is quicker and easier to use.

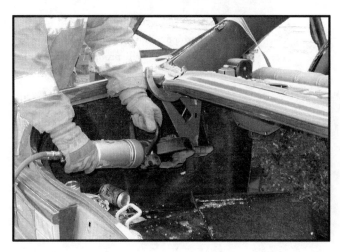

The rear deck is supported vertically by supports that extend downward in the trunk. They are cut low to provide as much clear space as possible.

Vehicle Extrication: A Practical Guide

Procedure 27 *(continued)*
Trailer Underride

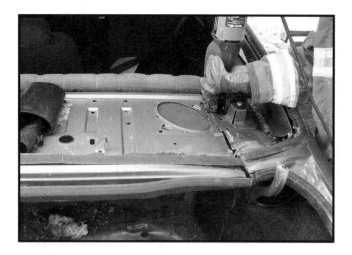

A reciprocating saw or air chisel is used to continue the rear deck cut to the rear seat area. After completing the cut on one side of the deck, the other side is completed.

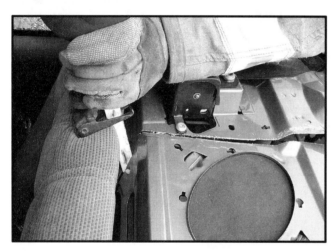

The cuts are extended as far as possible in the rear seat back supporting structure.

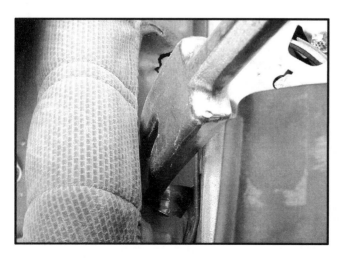

To continue the cuts downward, the rear seat cushion will have to be removed. A simple clip system will probably be used that can be exposed and separated with the use of a Halligan.

Procedure 27 *(continued)*
Trailer Underride

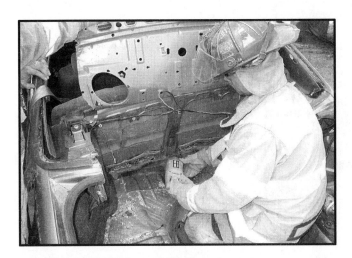

The rear deck may have a vertical support located in the center of the deck that extends downward. To improve access to the bottom of the vertical support, the rear deck can be folded up and out of the way.

The vertical support is cut as low as possible to reduce the size of the remaining structure. The smaller the remaining portion is the less chance there is of getting snagged or cut while working in that area.

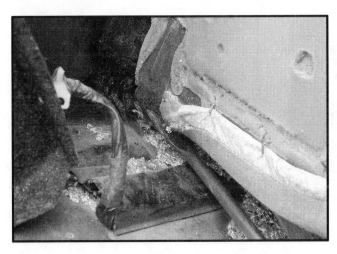

Any additional seat cushion attachment points can be cut with an air chisel or whichever tool fits in the space the best.

Vehicle Extrication: A Practical Guide

Procedure 27 *(continued)*
Trailer Underride

Once all the attachment points are severed, the rear seat cushion can be removed.

To increase the amount of vertical space in the occupant compartment, the lower rear seat cushion is removed. These are often held in place by spring clips that can be disconnected by pulling upward on the forward edge of the seat. In some vehicles, the seatbelts run through slots in the seat bottoms. The bulk of the buckle will prevent the easy removal of the seat cushion, requiring the cutting of the belts.

To access the front seat passengers, the front seat back needs to be moved out of the way. In two-door vehicles and vehicles equipped with bucket seats, a lever will control the position of the seat back. To lower the seat back, a handle is usually located on the outboard lower portion of the seat bottom. If the seat can not be lowered manually, the left and right seat back brackets can be cut away with hydraulic cutters.

Trucks and Tractor Trailers

Procedure 27 *(continued)*
Trailer Underride

Once the seat back is lowered or cut free, a long backboard can be positioned and the victim slid onto it. Before sliding the victim out, the seatbelt should be disconnected.

If the victim's legs appear to be trapped beneath the dashboard, the other seat back should be lowered to gain access to the area of entrapment. If the cutting tools can not be used for some reason, a come-along can be rigged to pull the seat back rearward.

Another option is to use the hydraulic spreaders to enlarge the space between the roof and the seat back.

Vehicle Extrication: A Practical Guide

Procedure 27 *(continued)*
Trailer Underride

Access to the area under and around the dashboard will be difficult, limiting the options available for freeing the victim's legs and feet. To move the dashboard up, hydraulic spreaders can be positioned under the dashboard for lifting. To reduce the chance of the floorboard collapsing, a piece of cribbing can be used under the lower spreader tip.

As the spreaders are opened, access to the pedals will allow the use of small pedal cutters and possibly full-size hydraulic cutters. When working in tight areas like the example shown, it would be prudent to have the valve on the power plant switched to the dump position to prevent accidental operation of the spreaders while working with other tools. In close quarters, it would not take much to inadvertently press against the control button of a hydraulic tool, allowing the tool to open or close in a manner that could trap a crewmember.

When the victim is ready to come out of the vehicle, straps can help move the victim onto the backboard while the interior crewmember maintains immobilization.

·10·

Extrication Procedures for School and Transit Buses

While the need for victim extrication from buses is rare, knowledge of practical tactical options can provide crews with the edge they need should they encounter a serious crash with multiple injuries and entrapments. Since buses are large, heavy, and designed with the occupant compartment above the typical impact area of most automobiles, bus occupants can usually avoid serious injury while occupants in the other vehicle may not fare as well.

Fig. 10–2 The height and mass of buses may expose occupants of automobiles to more risk of direct contact with the bus than is usually seen in automobile vs. automobile crashes.

Fig. 10–1 A comparison of the mass, height and built-in protection components of buses and automobiles reveals why the bus occupants often avoid serious injury.

If a bus is involved in a serious crash or rollover, the emergency exits at the rear and sides provide ambulatory occupants with a means of escape; they also provide good access for crews assigned to interior operations.

Vehicle Extrication: A Practical Guide

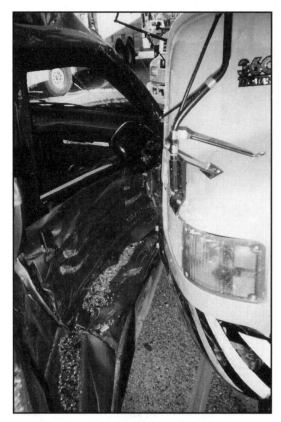

Fig. 10–3 The height and mass of buses may expose occupants of automobiles to more risk of direct contact with the bus than is usually seen in automobile vs. automobile crashes.

Fig. 10–4 Even after a severe crash, standard exit and emergency exit doors may remain operational, providing a means of egress for the occupants and access for fire and EMS crews.

If one exit is blocked or otherwise unusable, other exits may provide the needed opening to the interior. If a bus rolls over, or if all the exits are blocked, removing the windshield will create a large unobstructed area for crew entry and occupant removal.

Fig. 10–5 Side and rear doors may still provide a means of access and egress after a severe crisis.

Casual observation of any type of bus will reveal that a large portion of the passenger area is made up of windows, some of them doing double duty as emergency exits. When the doors are blocked or unusable, opening emergency exit windows, or simply removing a roof post may open up enough space for removing victims immobilized on backboards. The windows on some buses have a rubber hinge at the top of the window that can be cut with a knife and the window removed. Other types of windows can be opened and then pushed beyond their normal operating limits with a pike pole until the window disconnects from the rubber hinge and falls to the ground.

Extrication Procedures for School and Transit Buses

Fig. 10–6 Damage to one exit can demonstrate the importance of emergency exits for egress and access.

Some windows that are hinged on rubber gaskets can be opened and then pushed beyond their normal operating to completely release them from the vehicle. The window opening area and the area below the window should be kept clear of people as the window falls to the ground.

If access and egress from exits is blocked removal of the windshield can provide a large clear, yet cumbersome work area.

Other windows are attached with a rubber gasket and locking bead system that are easily removed by prying the bead out of the groove.

Fig. 10–7 One method of removing glass from a bus window is to pry out and remove the round bead from the groove in the gasket, making the gasket more flexible, allowing the glass to be pushed out of the gasket.

Once the locking bead is removed, the window can be pried out of the gasket. Another option is to cut through the inner part of the gasket with a razor knife to release the window.

Fig. 10–8 Another option for removing glass is to cut through the portion of the gasket that is holding the window in place. Once cut away, the glass can be pushed or pried outward.

Should those simple procedures prove inadequate, more complicated and time-consuming procedures may be needed.

Despite their strong construction, the roof and sidewalls of a typical school bus can be cut and removed with reciprocating saws or air chisels. Rivets identify the framework and reinforced areas, making them easy to avoid if necessary. Transit buses, on the other hand, have thick aluminum sidewalls that are difficult to penetrate, but their roofs can often be penetrated with simple tools. An air chisel, reciprocating saw, and even a flat-head ax and Halligan combination may be enough to remove a large section of roof.

Fig. 10–9 The roof on some transit buses is light enough that they can be easily opened up with hand tools. In this situation, a Halligan was used to strike a flat head axe that in turn cut through the lightweight metal of the roof structure.

Extrication Procedures for School and Transit Buses

Fig. 10–10 While the round roof emergency exit may provide egress and some access if a bus comes to rest on its side, large access holes may be needed.

When the simple procedures aren't adequate to extricate the victims of a crash, more aggressive procedures may be needed. The procedures in this chapter will address both the simple procedures and the more aggressive procedures that may be needed to free trapped victims from a serious bus crash.

PROCEDURES

Vehicle Extrication: A Practical Guide

Procedure 1
School Bus Emergency Exit
Side Door Exit

Emergency exits on school buses are simple in design and easy to operate.

To open emergency exit doors, the handles are are rotated one-quarter turn.

Once unlatched, the door is pulled open. A warning buzzer will be heard.

Rear doors are unobstructed, but this side door is blocked by a seat.

Any of the seat removal procedures can be used to clear the exit opening.

With the seat removed, personnel have easier access to the interior of the bus.

Extrication Procedures for School and Transit Buses

Procedure 2
School Bus Emergency Exit
Roof Hatch

For ease of operation, roof emergency exits are clearly marked and easily identified.

Like the emergency exit doors, a quarter turn handle is used.

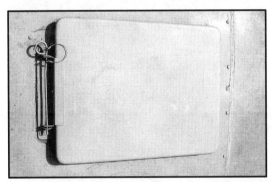

Some roof exits are equipped with devices that allow the hatch to be removed from outside.

To remove the hinge rod, the two clips are removed.

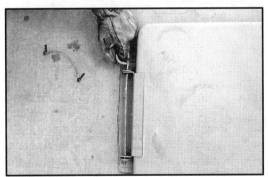

The hinge rod can then be pulled out of the hinge assembly.

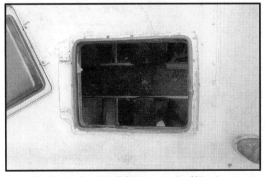

The hatch is then lifted up and off its base.

Vehicle Extrication: A Practical Guide

Procedure 3
Transit Bus Emergency Exit
Roof Hatch

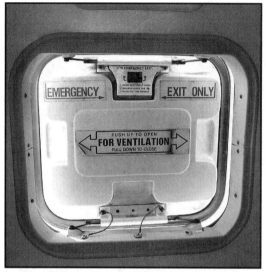

Roof hatches on transit buses serve as both vent openings and emergency exits.

To use as emergency exits, the vents are first pushed up to open.

The hatch release is then operated.

The roof hatch completely seperates from the body of the bus and vent mechanism.

Extrication Procedures for School and Transit Buses

Procedure 4
Transit Bus Emergency Exit
Window

Opening transit bus windows is a simple operation.

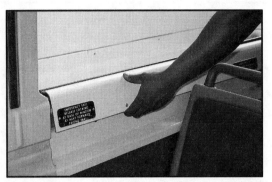

Directions for opening the window are marked on the operating mechanism.

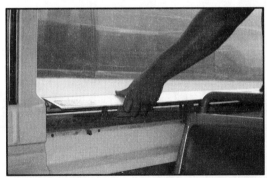

To release the window, the window base is lifted.

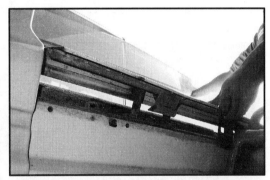

When lifted, the window is released from the latching mechanism.

The window is then manually pushed outward.

This is an exterior view of the latch.

Vehicle Extrication: A Practical Guide

Procedure 5
Removing Windshield Glass with Hand Tools

Windshield glass on a school bus can be removed with the use of a windshield saw or ax or by removing the rubber gasket and bead that holds the glass in place.

In this example, the pike of a Halligan is used to punch through the glass, and the center gasket is hooked. The glass and gasket are pulled outward to pull the rest of the glass out of the gasket and the gasket out of the windshield frame.

Once the gasket starts coming out of the frame, most of the glass can be removed in one large piece.

Extrication Procedures for School and Transit Buses

Procedure 6
Removing Side Window Glass with Hand Tools

Side window glass in older, second owner buses may be plate glass or may be tempered glass. This is evident in the side windows in this non-government owned bus. The upper window is plate glass, and the lower glass of the same window is tempered. The question then becomes *How do you know if the glass is tempered or plate?*

The answer to the question is that there is no way of telling, and that it does not make any difference. If a spring-loaded center punch is used on tempered glass, the entire piece of glass will break into small pieces when the punch is operated. If the glass does not break into small pieces immediately, but, instead, creates a spider web like the one in the upper piece of glass in the first photograph, it is plate glass.

If the glass is tempered, the window is cleaned out with the pike of a Halligan. If the glass is plate, it is also cleaned out with a Halligan or pike pole. The fact that the same tools are used regardless of the type of glass mounted in the frame makes knowing the type of glass unimportant. Both types of glass require the safe use of tools to avoid causing injuries to the bus occupants.

Vehicle Extrication: A Practical Guide

Procedure 7
Front Door Entry with Hand Tools

In situations when the driver of the school bus is unable to open the front door and the emergency exit doors are inaccessible, the glass can be removed from one of the front door windows to provide enough space for a crewmember to enter. The glass can be broken and the gasket removed to clear the window area or the bead can be pulled from the gasket. Once the bead is pulled, the window can be pried out of the gasket.

The larger of the two openings is selected if they are of different size. With the glass out of the way, a crewmember can slip through the opening.

This door opener handle is basic in design and serves the function of all manual openers: to provide enough leverage that the door can be opened and closed with the driver sitting and to lock the door in the closed position. While designs vary, common sense can be used to determine how the different devices operate.

490

Procedure 7 *(continued)*
Front Door Entry with Hand Tools

On this model of opener, a locking latch holds the door in the closed position and must be manually lifted to release the door.

The latch is lifted and the handle is moved out of the closed position.

As the handle is moved through its arc, the door opens.

Vehicle Extrication: A Practical Guide

Procedure 8
Front Door Entry with a Reciprocating Saw

If the front door of a school bus is jammed after a crash and won't operate normally, a reciprocating saw can be used to create an opening large enough to allow the exit of occupants from the bus. The first cut extends the opening from one lower window to the other.

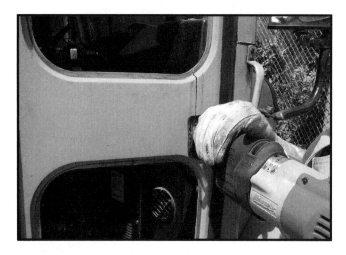

The divider between the upper and lower door windows are cut at the outer edge of the windows. After completing one divider, the other horizontal divider is cut in the same manner.

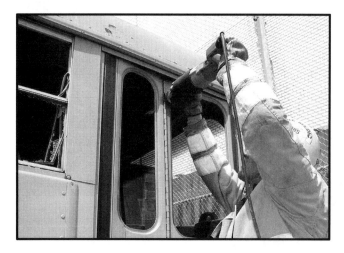

The vertical divider between the two upper windows is cut as high as possible to create the largest amount of space.

Procedure 8 *(continued)*
Front Door Entry with a Reciprocating Saw

The door opener connecting rod will prevent the center of the door from being removed and can be cut away with the saw.

To expedite the process, the cut is made in an area that contains the least amount of metal.

If the center of the door can be simply folded out of the way and there is enough space, cutting the connecting rod can be eliminated from the procedure.

Procedure 9
Front Door Entry with a Hydraulic Cutter

In the situation depicted in the photograph, the front door, even if operable, is blocked by another vehicle and can not be opened manually. Unlike the bifold door in the reciprocating saw procedure, this door assembly consists of two doors, each on its own hinge at the front and rear of the door opening. As with the other door, removal of the door starts with removing the glass.

The opening distance of most cutters will allow the blade to be wrapped around the center window divider to cut it with a single cut.

Some smaller cutters will require multiple cuts to remove the center window divider.

Procedure 9 *(continued)*
Front Door Entry with a Hydraulic Cutter

If the cutter blades won't open wide enough to wrap around the entire divider, multiple cuts can be made with the same results. A cut from one direction is made, severing the upper or lower half of the divider.

The cutter is then repositioned to cut the remaining divider. In this example, the upper cutter blade is repositioned in the first cut and the lower blade was positioned in the window opening to complete the cut. Another approach for the second cut, while more awkward, would require the cutter to be positioned low and the cut made upward.

There are two choices when severing the top of the door. As shown, the metal between the upper window and the upper edge of the door can be cut at the outer edge of the opening. This cut, in comparison to simply cutting the upper divider between the two upper windows, results in more clear space and fewer obstructions at head level.

Vehicle Extrication: A Practical Guide

Procedure 9 *(continued)*
Front Door Entry with a Hydraulic Cutter

When positioning the inside cutter blade, the tool operator should confirm that the cutter is in the desired position and not in an area of heavy reinforcement that could cause a fracture of the cutter blade.

To complete the removal of the forward half of this type door, the door opener connecting rod hardware or rod is cut.

The door can then be moved out of the way.

Extrication Procedures for School and Transit Buses

Procedure 9 *(continued)*
Front Door Entry with a Hydraulic Cutter

To remove the second door, the process is repeated starting with the severing of the upper window frame adjacent to the hinge.

The center window divider is cut.

The opening is completed, and there are no obstructions at head level.

Vehicle Extrication: A Practical Guide

Procedure 10
Emergency Exit Door Removal with an Air Chisel

When it is necessary to remove a rear or side emergency exit door at the hinge, an air chisel can be used with good results.

One solution to removing the door at the hinges is to cut away the fasteners that connect the door to the hinge pillar.

A sharp, curved chisel bit can be used to cut away the screw head, but the concave shape of the bit can make getting under the screw difficult. Once the chisel bit gets under the head of the screw, the shaft of the screw is cut easily.

Extrication Procedures for School and Transit Buses

Procedure 10 *(continued)*
Emergency Exit Door Removal with an Air Chisel

Another approach is to cut the shaft of the screw by positioning the curved cutter between the hinge and the door metal.

The ordinary or cold chisel type bit can get under the screw head easier than the curved cutter because of its straight cutting edge.

After all the hinge screws are cut, the hinge is moved out of the way. Initially, a Halligan can be used to fold the hinge open, but it is difficult to get a bite on such a small amount of metal. Once the hinge starts to open, a sledge hammer or the claw of the Halligan can be used to strike the hinge causing it to open.

Vehicle Extrication: A Practical Guide

Procedure 10 *(continued)*
Emergency Exit Door Removal with an Air Chisel

With the hinge out of the way and the gap between the door and hinge pillar exposed, the adz of the Halligan is driven into the gap and the door pried out of the opening. The door won't open completely until the door check is cut.

The door latch is a simple device and is easily pulled out of the receiver, allowing the door to be lifted out of the opening.

With the door completely removed, the size of the door opening is large and unobstructed.

Extrication Procedures for School and Transit Buses

Procedure 11
Popping an Emergency Door Open with Hydraulic Spreaders

Popping an emergency door open with hydraulic spreaders can be accomplished easily because of the simple design and construction of the door latch. The latch is located adjacent to the door handle; a location too high to attack from above as is done with automobiles. Instead, a Halligan can be used to create a purchase point for the spreaders below the latch. The spreaders can be initially operated at a comfortable height then worked upward in the gap between the door and the latch pillar.

If resistance is encountered after the door is popped open, the spreaders can be used to widen the door opening until the maximum opening distance of the spreaders is achieved.

The simplicity of the latch can be seen in this photograph taken after the door was popped open.

Procedure 12
Removing the Rear Window Posts with a Reciprocating Saw

If the clear area at the rear of a school bus needs to be expanded, a reciprocating saw can be used to remove one or both of the rear roof posts. To reduce the chance of the saw blade binding, the lower cut can be made first. By making the bottom cut first, there is less chance of downward pressure being exerted on the post as it is being cut.

If the blade starts to bind, the saw can be removed and the cut started again from another position. Even when starting from a position that will reduce the chance of the kerf closing as the material is cut, binding can occur. The binding caused by the closing of the kerf is a result of metal that is distorted and bent, allowing it to move as it is cut.

The next cut is made at the top of the post. With the bottom of the post cut, the lower kerf will allow the post to move a little, removing the chance of the upper cut binding.

Extrication Procedures for School and Transit Buses

Procedure 12 *(continued)*
Removing the Rear Window Posts with a Reciprocating Saw

With the second roof post and the partially cut door window frame remaining to be cut, it would be best to cut the roof post first instead of the door window frame. While this may be the ideal situation, the location of the tool operator may require the door window frame to be cut first.

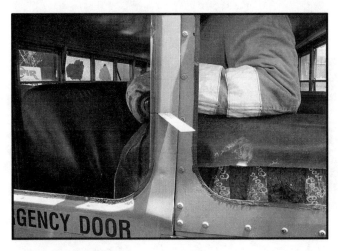

Cutting the roof post first would make it easier to cut the door window frame from the outside, an approach that would cause less binding of the blade. When cut in the manner shown in the photograph, the weight of the door window frame will cause the kerf to close as the cut proceeds. When cut from the outside, the weight of the window frame would cause the kerf to open up instead of close.

While the space in the rear window area has been opened up quite a bit, caution should be exercised when working around the remaining jagged roof posts.

Vehicle Extrication: A Practical Guide

Procedure 13
Removing the Rear Window Posts with a Hydraulic Cutter

When additional space is needed and a roof post needs to be removed to create the space, hydraulic cutters are the best choice of tools. The advantage of the cutter is the ability to sever the heavy steel used in the construction of the post with little chance of binding. If the roof post is bent, some parts of the post will be in tension and some in compression, which can lead to binding of a reciprocating saw and to a lesser extent, an air chisel. To cut the roof post with hydraulic cutters, the bottom post is cut first if possible.

By making the upper cut after the bottom cut, the falling post is less likely to strike the tool operator.

The removal of the roof post may be a good option when more room is required for access and removal of victims and when the door is damaged and can not be opened.

Extrication Procedures for School and Transit Buses

Procedure 14
Rear Wall Removal with Hydraulic Tools

When it is necessary to create more space than can be provided by opening the rear door or removal of a roof post, the rear wall can be cut and moved out of the way with hydraulic tools. The procedure starts by separating the rear wall from the roof by cutting high on the roof post.

To create a hinge or crease, the strength of the upper wall must be weakened with a relief cut. The cut will sever the reinforced areas around the window, making it easier to fold to the side.

To fold properly, the bottom edge of the wall needs to be separated at floor level. To start the process, hydraulic cutters are used to make a cut at floor level from the door opening.

Vehicle Extrication: A Practical Guide

Procedure 14 *(continued)*
Rear Wall Removal with Hydraulic Tools

The cutters are then repositioned to extend the first cut. The process of extending the cut is continued as far as possible.

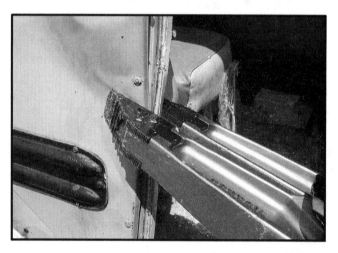

The wall may be difficult to grasp and pull out by hand, but the spreaders can be used to create a good grip and extra leverage. After opening the spreader a few inches, close the spreader tips and clamp them down on the wall. It is important to close the spreader tips firmly to prevent the tips from slipping off the wall as it is pulled around.

The spreaders are gripped firmly by the handles and the tool is walked around, causing the metal to crease and fold. For more work space, one of the seats can be removed.

Extrication Procedures for School and Transit Buses

Procedure 14 (continued)
Rear Wall Removal with Hydraulic Tools

An additional option starts by making the same type of relief cut in the rear wall.

Another relief cut is made at floor level.

The spreaders are then used to push the wall away, causing the relief cut to tear open as the spreaders open.

Vehicle Extrication: A Practical Guide

Procedure 15
Roof Entry with a Reciprocating Saw

To gain access to victims inside a school bus through the roof, a three-sided flap or four-sided removal can be performed with a reciprocating saw. To avoid delays and wear on the saw blade, the reinforced areas of the roof are avoided when possible by cutting next to (but not through) areas containing lines of rivets. To create a starter hole for the saw blade, make a hole by positioning the pike of a Halligan against the metal then striking the Halligan with an ax or sledge hammer. When preparing to make the hole, keep the area on the inside of the bus clear of occupants.

After placing the saw blade in the hole, start the cut and continue downward to the bottom edge of the roof.

A second starter hole is made that will serve as the starting point for the second downward cut. The positioning of the second starter hole will allow a cut to be made between the two holes without having to cut through any reinforced areas.

Extrication Procedures for School and Transit Buses

Procedure 15 *(continued)*
Roof Entry with a Reciprocating Saw

The order of the second and third cuts is unimportant and, in this example, the second cut connects the starter holes.

The third cut extends downward from the starter hole to the same level where the first cut ended. The location of these two ending points will determine where the fold in the roof will be made. To lower the location of the fold, making it easier to get in and out of the hole, the cuts should be continued downward.

With the three cuts completed, the roof can be folded down. Another option is to make a fourth cut that totally eliminates the flapped metal. The downside to removing the flap is the creation of another sharp edge that will have to be avoided or covered with an EMS blanket or salvage cover.

Procedure 16
Roof Entry with an Air Chisel

A three-sided roof flap can be performed with an air chisel or reciprocating saw, but the air chisel is less likely to bind up when working on a deformed roof. When a reciprocating saw is used on a bent roof, there is a chance of the kerf closing as the cut is made, leading to a binding of the blade. An air chisel makes a wider kerf and has more ability to power through the binds without getting jammed. Three cuts are made to flap the roof with the sequence of cuts unimportant. To avoid additional work resulting in limited benefit, the cuts are made just to the inside of the reinforced areas that can be identified by a vertical row of rivets.

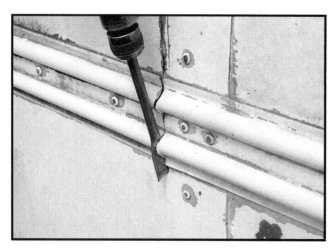

A T or dual cutter can be used on the flat sections of the roof, but a curved cutter bit will be needed in areas of multiple layers of metal.

After the third cut is made, the flap can be pulled out and folded down. To eliminate the presence of the flap in front of the opening, a fourth cut can be made to remove it.

Extrication Procedures for School and Transit Buses

Procedure 16 *(continued)*
Roof Entry with an Air Chisel

Unlike the reciprocating saw with a long blade that will penetrate both the roof and the ceiling, the air chisel will only cut one layer at a time. To flap the inner layer of metal, the procedure is the same as when cutting the outer layer except that the cuts are made a couple of inches in from the original cuts. This positioning gives the tool operator a little bit of room to work and manipulate the tool while still creating a hole large enough to be useful.

As with the outside cut, three cuts are made if the metal is to be bent down, four cuts are made to completely remove it.

As the third cut is completed, the inner layer is bent down and outward. Bending the roof into the bus would create an obstruction in the interior and should be avoided. If the flap won't bend outward, a fourth cut can be made, eliminating the flap completely.

Procedure 17
Side Wall Removal with a Reciprocating Saw

When a section of the side wall of a school bus must be removed with a reciprocating saw, the process can be made easier if the initial cuts through the top of the wall are made with hydraulic cutters. The initial cut can be made with a reciprocating saw, but it will take longer and will require more replacement blades. To maximize the space created by the removal of the side wall, start the cut is directly behind one of the seats.

The thickness of the metal at the top of the wall can be seen as the hydraulic cutter completes the cut.

To extend the cut, position the saw blade at the bottom of the cut made by the cutters.

Extrication Procedures for School and Transit Buses

Procedure 17 *(continued)*
Side Wall Removal with a Reciprocating Saw

The multiple layers of metal at the top of the wall can be seen in this photograph taken after the sidewall was removed.

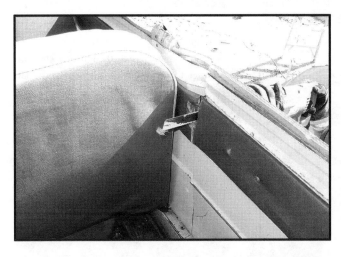

The advantage of making the cut behind the seat back can be seen in this photograph. If the cut were started in front of the seat, the saw blade would strike the seat bottom, preventing the cut from extending to the floor.

The cut is extended downward to the level of the floor.

Vehicle Extrication: A Practical Guide

Procedure 17 *(continued)*
Side Wall Removal with a Reciprocating Saw

If the level of the floor can not be approximated, the tool operator can determine the location of the floor by the resistance felt as the blade makes contact with it.

There are a couple options for creating a starting point for the horizontal cut. One option is to punch a hole though the metal with a sledge hammer and the pike of a Halligan; the other requires the outer layer of metal to be folded back with a Halligan.

By prying with the adz or the fork of the Halligan, you can tear the metal away, creating a corner that can be folded up.

Extrication Procedures for School and Transit Buses

Procedure 17 (continued)
Side Wall Removal with a Reciprocating Saw

With a little manipulation of the Halligan, the inner layer of metal can be bent enough that the saw and blade can be positioned horizontally and the cut started.

The cut is extended along the floor toward the other vertical cut. If resistance is encountered while the cut is being made, the tool operator should determine if the resistance is caused by a vertical structural support or because the saw blade is striking the floor. If the floor is causing the resistance, the saw can be backed up a few inches and a new cut made above the floor level.

As the tool operator approaches the vertical cut, adjustments are made to ensure that the horizontal and vertical cuts meet.

Vehicle Extrication: A Practical Guide

Procedure 17 *(continued)*
Side Wall Removal with a Reciprocating Saw

This view from above shows the size and the shape of the vertical supporting members.

As the cut is completed, the side is lifted out.

This example of a side wall removal spanned a distance of two windows, and a seat was removed to create a clear area for movement of equipment and personnel.

Extrication Procedures for School and Transit Buses

Procedure 18
Side Entry with an Air Chisel

If the side of a school bus is to be removed, it is important to size up the area and decide how big the hole needs to be to accommodate the movement of personnel, equipment, and victims. Once that decision is made, the glass that will be affected by the operation can be removed. Steps should be taken to protect any occupants that are in the area where the glass is being removed.

In this example, the metal between two window posts is to be removed. The sequence of cuts is unimportant, and, in this example, one of the vertical cuts is being made first. In the hands of an experienced operator, the curved cutter is the best choice of air chisel bits because it can be used for virtually all shapes of metal encountered on a bus. A T or dual cutter can be used in flat, smooth areas, but the curved cutter will be needed in the shaped and reinforced areas.

To reduce the chance of the air chisel bit becoming stuck in the kerf as the metal settles downward as it is cut, the bottom cut is made next followed by the second vertical cut.

Vehicle Extrication: A Practical Guide

Procedure 18 *(continued)*
Side Entry with an Air Chisel

If the rub rail is a problem, it can be removed by cutting away the rivets and screws then prying or cutting it away from the underlying panel.

The area at the top of the wall (beneath the bottom of the window) is reinforced and can be difficult to cut completely if unexposed. The fourth cut results in the outer layer of metal being removed and exposing more of the supporting structure, making it easier to cut.

After removing the outer layer of metal, it becomes a matter of cutting away large sections of metal as they are encountered. Again, the curved cutter is the most effective bit for this type of operation.

Extrication Procedures for School and Transit Buses

Procedure 18 *(continued)*
Side Entry with an Air Chisel

With the outer and middle layers of metal out of the way, the interior wall section can be cut away in the same manner as the exterior wall.

After being cut, the interior wall section can be folded down or totally removed. In this example, the inner wall was folded down out of the way.

To complete the opening, the window frame in the center of the window can be removed by striking it with the bar of the Halligan or by cutting it away with the air chisel.

Vehicle Extrication: A Practical Guide

Procedure 19
Removing a Steering Ring with Hydraulic Cutters

If a collision has resulted in the driver being trapped between the seat back and the steering ring, space can be created by using hydraulic cutters to remove the spokes and ring or parts of the ring only. Small specialty cutters or full-size cutters can be used to cut through each one of the spokes.

Another option is to cut the steering ring in two locations adjacent to the spokes.

In this example, the second cut is being made on the opposite side of the ring in an area that may be obstructed by the victim or dashboard. If the cutter can not be positioned properly because of obstructions, it may be possible to make one cut then forcibly turn the steering wheel to expose the part of the ring needed for the second cut.

Extrication Procedures for School and Transit Buses

Procedure 20
Relocating a Steering Column with a Come-along

When cutting the steering ring away does not create enough space to treat or remove the driver, moving the entire steering column and wheel may be a good option. The procedure is basically the same as the procedure used on an automobile. As with autos, this procedure should be considered a last choice because of the potential danger of the column breaking loose and striking the victim. The first step in rigging the steering column is to pass the master link up and through the lower portion of the steering wheel and across the spokes.

The master link is then passed through the top opening in the wheel and attached to the block of a come-along that is positioned on the hood of the bus, just forward of a slider crib. The slider crib will help keep the chain out of the dashboard and moving smoothly.

The steering column is wrapped low with the chain to provide a good, secure anchor. The most effective way to make the wrap on the steering column is to grasp the hook at the end of the chain and pass it over the column. The hook is then grabbed by the other hand and pulled, and all the slack chain removed.

Vehicle Extrication: A Practical Guide

Procedure 20 *(continued)*
Relocating a Steering Column with a Come-along

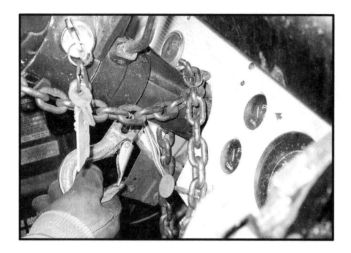

The hook is passed over the column again, completing the wrap.

Next, the hook is passed through the lower opening of the steering wheel then through the upper opening of the steering wheel, and all the slack chain is pulled forward onto the hood.

The chain shortener is attached to the chain to remove all the slack chain. When the wrap is finished, the left and right sides of the steering wheel should look like mirror images of one another.

Extrication Procedures for School and Transit Buses

Procedure 20 *(continued)*
Relocating a Steering Column with a Come-along

The outside crewmember responsible for setting the anchor chain and operating the come-along walks to the side of the bus, grasps the come-along, and pays out cable until the come-along is positioned at the front of the bus. The come-along should be positioned so the cable pays out straight and unobstructed. To resist downward crushing when pressure is applied to the come-along, ladder cribbing or single pieces of cribbing are positioned on the hood to distribute the load.

The anchor chain is attached to a strong member of the bus chassis.

When a sling hook is used to choke the chain, latches are helpful in keeping the chain in the hook during the rigging and set-up. In this example, the latch has sprung and serves no function. This photograph serves to emphasize the importance of checking the chains, hooks, and come-along after the slack is removed but before the system is put under load. While the broken latch does not change anything in this example, in other situations the broken latch could allow the chain to fall out of position.

Procedure 20 *(continued)*
Relocating a Steering Column with a Come-along

Before operating the come-along, the officer calls for the crewmembers to check the rigging. Once checked, the come-along is operated.

As the steering column moves upward, the master link moves forward on the slider crib. When the top 4x4 reaches the forward end of the lower 4x4s, additional 4x4s can be added to extend the range of travel.

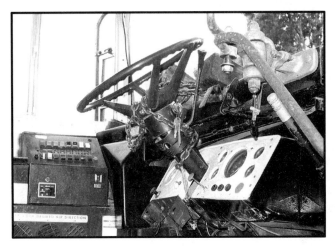

The operation is continued until the steering column has moved enough to access and remove the driver.

Extrication Procedures for School and Transit Buses

Procedure 21
Driver's Seat Back Removal with Hydraulic Tools

In situations when the dashboard cannot be moved forward, removing the seat back may provide the space needed to treat and remove the victim. To create the desired space, not only will the seat back have to be removed, but the divider behind the seat should be removed to simplify the entire process.

The tool operator should size up the construction and type of attachment of the divider before starting the cut. Cutting into metal blindly can result in the tips of the cutter being placed in a position that causes them to pierce the metal, a situation that can cause the cutter blades to become overloaded and fracture.

In this example, the method of attachment has been determined and the cutters positioned to make the cut.

Procedure 21 *(continued)*
Driver's Seat Back Removal with Hydraulic Tools

After cutting through the vertical attachment structure, the divider is pushed out of the way to expose both sides of the driver's seat back.

The construction of the divider can be seen in this photograph.

If necessary, the horizontal member of the divider can also be cut out of the way.

Extrication Procedures for School and Transit Buses

Procedure 21 *(continued)*
Driver's Seat Back Removal with Hydraulic Tools

The construction of the seat back is examined and the cuts are made to sever it from the seat base.

To expose the other side of the seat, the seat back can be forced upward.

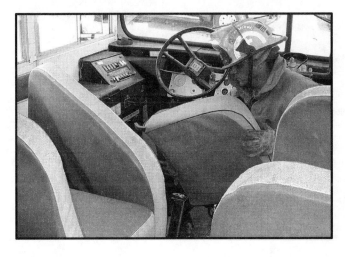

The remaining attachment points are severed and the seat back removed.

Procedure 22
Seat Post Removal with a Reciprocating Saw

When removing seats from a school bus, a reciprocating saw can be used to cut the legs but is not useful for cutting the seat frame away from the bus wall. When cutting the legs, position the saw close to the floor to remove as much of the leg as possible.

A short metal cutting blade works well for this procedure because it is less likely to strike the floor than the long blade.

The remaining leg and base can create a trip hazard, making the use of the reciprocating saw a good backup tool but not a good primary tool.

Procedure 23
Seat Removal with an Air Chisel

If a seat needs to be removed from a school bus, an air chisel can be used to cut both the legs and the wall mounting brackets. To avoid having the bit become jammed, the curved chisel bit is used to cut around the perimeter of the leg (not through the center).

By cutting around the perimeter, half of the blade can be kept outside the cut, reducing the chances of a bind.

When the cut is completed, the top of the foot is flat and without any sharp edges.

Vehicle Extrication: A Practical Guide

Procedure 23 *(continued)*
Seat Removal with an Air Chisel

Another approach that takes a little longer than cutting the leg results in removal of the leg and foot. A cold chisel bit is used to get under the head of the bolt or screw and to shear it off.

The process is completed on all the bolts and screws that attach the foot to the floor.

After shearing off all the heads, the foot can be removed completely, leaving no obstructions or trip hazards.

Extrication Procedures for School and Transit Buses

Procedure 23 *(continued)*
Seat Removal with an Air Chisel

To gain access to the wall-mounting brackets, the seat bottom is removed by lifting or prying away from the seat frame. With the seat cushion out of the way, the method of mounting against the wall can be evaluated.

In this example, the heads are sheared off, but this time a curved cutter is used. The cold chisel bit is a better choice for this operation, but in this example, the curved cutter was able to be positioned completely under the head of the fastener.

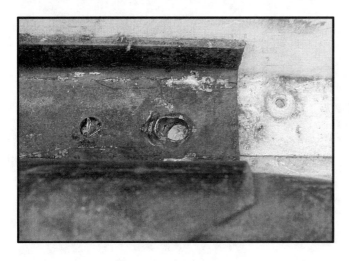

As with the foot attachments, once the head is sheared off the bolt or screw, the frame is simply lifted off the attachment point.

Vehicle Extrication: A Practical Guide

Procedure 24
Seat Removal with Hydraulic Tools

Hydraulic spreaders and cutters can both be used to remove bus seats with equal effectiveness. When spreaders are used, one tip is placed against the floor and the other against the seat frame.

The floors in school buses are usually stout and do not require any cribbing to prevent the tips from penetrating the floor. As the spreaders are opened, the screws attaching the seat to the floor will tear out of the floor suddenly and with considerable force. To avoid injuries, the area around the seat should be kept clear of personnel during the procedure.

In this example, one screw tore out of the floor and the other tore through the base of the leg. If the second leg base does not tear out of the floor, the spreader is repositioned and the procedure repeated.

532

Extrication Procedures for School and Transit Buses

Procedure 24 *(continued)*
Seat Removal with Hydraulic Tools

After breaking away the seat legs, the seat frame can be broken away from the wall. The spreader tips are positioned close to the wall to concentrate the spreading force on the wall attachment point and to avoid lifting the seat and frame.

Another approach, while not as simple as the spreaders, involves cutting the seat away with hydraulic cutters. The legs are cut first so the seat can be pivoted upward to gain access to the wall mounting hardware.

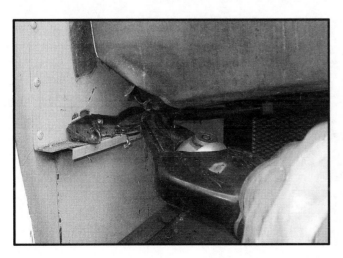

The wall-mounting hardware is sized up and the cutters carefully positioned before making the cuts. When cutting heavy steel like the steel used in the seat-mounting hardware, it is best to wrap the cutter blades completely around the object before starting the cut. If the cutter tips are not wrapped around the object, but instead have the tips positioned to punch through the metal, there is a greater chance of fracturing the cutter blades.

Vehicle Extrication: A Practical Guide

Procedure 25
Lifting a Bus with Air Bags

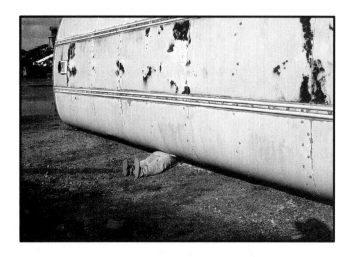

Lifting a bus that has rolled over may at first may seem like an impossible task, but after considering the amount of weight actually being lifted, the task seems less insurmountable. When sizing up the lift, crewmembers should assess both the outside and inside of the bus.

If the windows are open or broken, the crewmember can carefully walk from one opening to the next to avoid adding load to the bus.

When sizing up the interior, the crewmembers assigned to the inside should avoid walking on the wall, roof, and roof posts of the bus to avoid adding additional weight to the vehicle and the victim trapped beneath it. While walking on the roof and walls may seem like a safer approach—which it is for the crewmember—the victim could certainly suffer from the added weight.

Extrication Procedures for School and Transit Buses

Procedure 25 *(continued)*
Lifting a Bus with Air Bags

The locations of the hidden structural members are easy to identify by simply looking at the surface of the metal. Common sense would indicate that rivets or some other type of fastener are used when two or more pieces of metal are to be joined. By applying that knowledge to this roof, you can identify the location of the structural strength of the roof by the presence of metal fasteners. For the greatest amount of strength, cribbing and/or lifting tools are used directly inline with the fasteners.

After selecting the best lifting points, put the air bags into position. Cribbing is inserted to maintain the existing space and to prevent any downward movement of the bus once the lift is started. To maximize the lifting potential in terms of weight, the air bags are not stacked (series lift), but instead each air bag is positioned between the ground and the bus (parallel lift).

If lifting height is more important than lifting capacity in weight, the bags can be stacked.

535

Procedure 25 *(continued)*
Lifting a Bus with Air Bags

Whether the air bags are stacked (series) or side by side (parallel), the crewmembers positioning them from the outside should remember to check the interior of the bus for best placement. What may look like a good, sturdy position from the outside may actually place the bags under a piece of glass or in an open window. This task can be accomplished with the assistance from the interior crewmember.

Good teamwork is needed if the lift is to be made smoothly and safely. The crewmembers cribbing the bus as it is lifted will need enough cribbing nearby to avoid delaying the operation. Under the best circumstances, the air bag controller operator has a clear view of the operation and is able to see and stop a problem quickly. The officer should closely coordinate the lift and decide when to stop it.

As the lift proceeds, cribbing is built to support the bus in the unlikely chance that one of the air bags were to fail (rupture). Careful observation of the victim will reveal when the victim is free of the bus. Once free, the inside and outside crewmembers can work together to slide the victim from beneath the bus.

Extrication Procedures for School and Transit Buses

Procedure 26
Lifting a Bus with Hydraulic Spreaders

Any lifting procedure will benefit from a good size-up that will clearly identify the type and scope of the entrapment. During the size-up the crew can identify any complications that may occur during the lift, along with solutions to minimize their effect. Interior crewmembers may be able to size up the entrapment and provide victim assessment and treatment while the tools are being set up.

When positioning a spreader tool for a lift, the potential lift areas should be quickly examined for structural integrity and reinforcement. In this example, the vertical line of rivets above the spreader is an indication of reinforcement, making it a good choice as a lift point. Before starting the lift, cribbing is put in place to prevent the bus from further settling. On soft ground, cribbing or other type ground pad will be needed to prevent the spreaders from sinking into the earth.

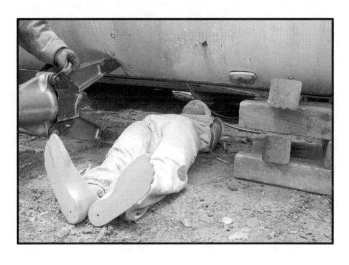

As the lift is started, the crewmembers responsible for building the cribbing are going to have to hustle to keep up with the rapid lift of the bus. While the spreader operator may be tempted to lift quickly and pull the victim out, the downside of shifting loads, unanticipated entanglement, and tool failure should be considered. The location of the spreader tips that may have looked good at the beginning of the lift may not look so good midway through the operation. Should the tips slip out of position, the time spent building the box crib would be recognized as time well spent.

· 11 ·

Sport and Race Vehicles

Entire books could be written about race car fire suppression and extrication procedures. With so many different types of vehicles and procedures, production of any comprehensive guide would be impractical, if not impossible. This book guides the reader in learning how to obtain the information needed to work safely in the racing environment.

Working at a racetrack is different from working on the streets outside of the racetrack. On the street, most firefighters and EMS personnel are accustomed to taking control of the scene, redirecting traffic, and basically calling the shots. In the racing world, fire suppression, extrication, and medical crews are part of a larger controlling organization. As the size and prominence of the track and sanctioning body increases, so does the degree of control and coordination by the event operations personnel at the top of the organization.

Professionally operated events have written and unwritten policies and procedures that govern the operation of emergency service crews. The degree of control exercised, and the expectation of compliance is necessary because of the environment of the racetrack. Crew members working at a crash scene on a track may have to respond to incidents while other race cars not involved in the crash continue making laps, instead of being rerouted as would happen outside of the racetrack. The importance of learning and understanding the role and responsibilities of an on-track technician can't be overstated.

To prepare local fire department personnel for working at large events, some sanctioning bodies conduct policy and procedure training before the event. This type of training relieves some of the apprehension associated with working at a major event and allows the local workers to meet and learn from full-time racing personnel. At pre-event meetings and training sessions, local track crews have an opportunity to learn about the vehicles involved in the event, the equipment available, track policies, response criteria, and method of dispatch.

It's important to understand that the staff of major events are usually full-time veterans of the sport and are going to be the best source of information and direction. At first, firefighters and officers may have a hard time understanding why they can't respond immediately to a crash that occurs directly in front of them, but by listening and learning from the professionals, injuries to response personnel can be prevented.

Construction Basics

While there is a great deal of difference between standard automobiles and racing vehicles, there are some basics that do cross over from the street to the track. In fact, some vehicles found on the quarter-mile drag strip may have been driven to the track by the owner. On the other hand, high-performance race cars are brought to the track in large trailers that serve as both a mode of transportation and a shop. Whichever type of vehicle is encountered, there are a few basics that apply to all types of race cars.

Roll bars and cages

Some race cars actually are standard automobiles or trucks that are modified to provide the needed characteristics and equipment required of the sport. Street cars that race at the local drag strip on a Friday night may not provide many surprises for the crew assigned to a crash. The vehicle may be completely stock, or the vehicle may have upgrades to reduce weight or increase safety. One method of increasing safety in all types of racing vehicles is the roll cage. Roll cages are designed to provide the driver with protection in case the vehicle rolls or is involved in a crash. In drag racing, the lower the vehicle's time in the quarter mile race, the more pieces are required in the construction of the roll cage. In all types of race cars, careful planning is needed when cutting a roll cage to free a driver. Some sanctioning bodies use a procedure that specifies where and when each member of the roll cage is cut. Using these predetermined cuts reduces the chance of unnecessary and missing cuts.

Two materials used in roll cage construction are mild steel and chrome moly. Chrome moly is a metal made of chromium and molybdenum, along with other elements; it is used for its strength and weight characteristics. If pieces of chrome moly pipe and mild steel pipe were placed next to each other, most people wouldn't be able to tell the difference between the two. Trying to differentiate between the two after the cage is painted is almost impossible. While the name chrome moly sounds high-tech, the good news is that a reciprocating saw can be used to cut both mild steel and chrome moly with no discernable difference. With a sharp blade, mild steel and chrome moly pipe used for roll cages can both be cut in seconds. Hydraulic cutters can also be used to cut the pipe. While chrome moly can be cut easily, manufacturers of roll cages state that cutting in the area of a weld may be more difficult because of changes in the metal caused by welding.

Fuel and fuel tanks

When working at an event, crew members should make a point to learn about the type of fuel and fuel tanks used on the vehicles at the

Sport and Race Vehicles

event. Knowledge of the types of fuel is important because some fuels aren't visible when burning. This means a crew could walk into a puddle of burning fuel and not be aware of it until it's too late. If a driver exits a race car and is slapping at his legs, he's probably burning or been exposed to fuel that was burning in the car.

Fuel tanks in street cars may be different in construction and location from those used in high-performance race cars. Some fuel tanks are located in the front of the engine, taking advantage of the g-forces associated with acceleration. Should a vehicle roll over and come to rest on its roof or side, a check valve may prevent fuel from leaking out of the fuel tank vent.

Steering wheels

Removable steering wheels are installed in some racing vehicles to make it easier for the driver to get in and out of the seat. Extrication and medical crews can take advantage of this feature, making it easier to remove an injured driver from the vehicle. The design and operation of the removable steering wheels is simple and intuitive, but the crew members should determine which type of steering wheel is being used before the race begins.

Seats and belts

The design and construction of seats, belts, and other restraining devices is based on the type of vehicle and the requirements of the sanctioning body. The belt systems used to restrain the driver may use multiple attachment points to adequately secure the driver. The operation of the more complex belt systems is different than those used in automobiles.

Doors and other means of access to the interior

Race cars that resemble street cars may be equipped with a lightweight door that opens differently than a standard automobile door. If no outside door handle is found, a hole in the door window may be found that provides access to the latch lever mounted on the inside of the door. If it's necessary to remove this type of door at the hinges, a quick examination of the hinge may provide the clues needed to remove it without any cutting.

Other cars may look like they have doors, but only have the image of a door painted on the side of the vehicle. If the vehicle is built with a window opening without glass or Plexiglas, the driver may use the opening to slide in and out of the car. Crews responsible for working on cars with this arrangement should determine how the driver will be accessed and removed from the vehicle if he needs to be immobilized. Monster trucks use another type of arrangement for the driver to enter and exit the vehicle. Instead of climbing through an open window, the driver climbs up through the bottom of vehicle's roll cage to reach the driver's seat. To keep dirt out of the driver's compartment, a Plexiglas divider is used to cover the opening at floor level.

Funny cars, one of the most popular types of drag racing vehicles, have bodies that are hinged at the rear to provide access to the driver's seat and engine. To raise or lower the body, a simple latch system at the front corner is operated. During the time between runs, the body is held up by supports, but those supports are kept in the pit area and not attached to the vehicle. If an incident requires that the body of the car be

Vehicle Extrication: A Practical Guide

completely removed, the locking pins at the rear of the car can be removed, releasing the body from the chassis. If necessary, when time is a factor but expense isn't, the body can be pushed up and rearward until it's is out of the way. Pushing the body over the top and beyond its normal operating limits will result in damage to the body and/or framework.

Shutting down the engine

Some vehicles are equipped with engine shut-off switches located at the rear of the vehicle. These switches allow an emergency crew member to shut down the electrical system of the vehicle without having access to the interior of the vehicle. The switches are simple in design and appearance, making them easy to understand and operate.

If the vehicle isn't equipped with a shut-down switch, the correct switch will have to be located in the driver's compartment. By the nature of the sport, monster trucks are more prone to roll-over situations than most other types of racing vehicles, making the application of a remote control shut-down a valuable tool. If the monster truck driver is unable to turn off the vehicle's engine, a track official can transmit a signal to the vehicle's ignition system to shut it down.

SUMMARY

Some extrication procedures used on the street can be adapted to racing vehicles, while other procedures used in racing are specific to the types of vehicles being used. Standard extrication tools can be used on roll cages with good results. Response policies and procedures are quite different at the racetrack than on the street, making planning and training an important component of response readiness.

Appendix:
Things to Think About

Vehicle Extrication: A Practical Guide

This oval track simulator gives local firefighters an opportunity to train prior to an event.

As with other types of extrication work, initial contact with the driver is important in size-up.

Race cars often require the removal of the roll cage prior to removing the driver.

Hydraulic cutters or a reciprocating saw can be used to cut cage both mild steel and chrome moly.

Cross-section of a roll cage cut with a hydraulic cutter.

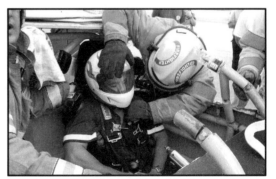

During training, crews can learn how to release belts and harnesses.

Fig. A–1 Pre-event training

Sport and Race Vehicles

Instead of exterior handles, some types of doors have levers that are operated from inside.

Interior view of a door lever.

Some drag racers use lightweight replacement doors instead of the heavier originals.

This lightweight door hinge is simple in design and easily disassembled after removing a clip.

Once the door is open and removed, roll cage members can be cut out of the way.

This cage creates more of an obstruction because of the additional cage member.

Fig. A–2 Drag racing doors

Vehicle Extrication: A Practical Guide

This Monster Truck has a battery kill switch handle located beneath the rear bumper.

Some switches aren't as visible as others.

To operate the switch, the ring is pulled outward.

When activated, a simple switch opens, disconnecting the battery.

Some drag racing cars also use a switch at the rear of the vehicle.

All the switches are simple in design and easy to operate.

Fig. A–3 Remote electrical shutdown switches

Sport and Race Vehicles

Nets are used to keep driver's heads inside the vehicle during a crash.

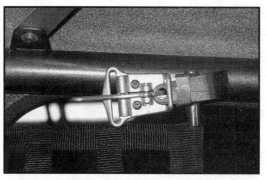

A simple latch holds the net in place.

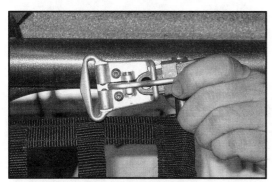

To release the net, the lever is moved to the open position.

The net and rod are released and can be lowered out of the way.

To remove the steering wheel, the collar on the back of the wheel is squeezed toward the driver.

While squeezing the collar against the steering wheel, the wheel is pulled off the spline.

Fig. A–4 Interior obstructions

Vehicle Extrication: A Practical Guide

Drag racing's funny cars don't use doors for the driver to enter and exit the vehicle.

Instead a latch is used at the front of the vehicle.

This interior view of the latch from below shows the latch in the locked position.

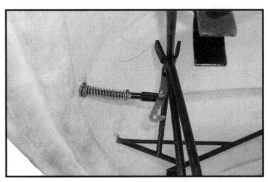

When the T handle on the outside is pulled, the latch releases from the striker.

The rear of the body is hinged on a removable locking pin.

With the front latch released, the body can be lifted and held in place with a stand.

Fig. A–5 Gaining access in a funny car

Sport and Race Vehicles

The funny car roll cage creates a snug fit for the driver.

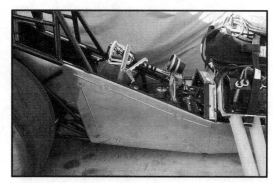

The driver's feet can be accessed by removing the panels on the side.

Rear view of roll cage.

Side view of roll cage.

Pre-planning procedures for vehicles like this dragster will save time when it's needed most.

A view of another style dragster with the side panels removed.

Fig. A–6 The funny car and dragster driver compartments

Vehicle Extrication: A Practical Guide

Like other race cars, the funny car utilizes a removable steering wheel.

To release the steering wheel, the collar is pulled toward the driver.

While squeezing the collar, the steering wheel is pulled off the column spline.

Understanding how belt systems release is an and important part of training and pre-planning.

This accelerator pedal has a clip that grips the end of the driver's boot.

Driver's foot in position.

Fig. A–7 The funny car driver compartment

Sport and Race Vehicles

To enter the driver's compartment, the driver climbs up from between the front and rear wheel.

Looking up between the body and the frame at the driver's floor hatch door.

The hatch door is hinged and is held closed with Velcro (View from driver compartment).

Like other racing vehicles, Monster Trucks use removable steering wheels.

To release the steering wheel, the locking pin is pressed in toward the column.

The steering wheel is then pulled off the shaft.

Fig. A–8 Primary access to the monster truck driver compartment

Vehicle Extrication: A Practical Guide

With the removable bed panel removed, the rear of the driver's compartment can be seen.

Plexiglas or another type of clear plastic separates the engine and driver's compartment.

In addition to the divider, the roll cage obstructs access to the driver's compartment.

A different roll cage design.

The front of the driver's compartment can be seen through the front wheel well.

Like the rear of the driver's compartment, clear plastic separates it from the front end.

Fig. A–9 The monster truck driver compartment

Some body panels are attached with panel buttons.

To release the panel, a screw driver or specialized tool is used to turn the button one-quarter turn.

Locking pins are used to attach other panels.

To remove the pin, the button at the end of locking pin is depressed.

Other panels are bolted into position.

Fenders are lightly attached for easy replacement.

Fig. A–10 Removing monster truck body panels

Index

4x4 post (vehicle stabilization), 129

A

Access means (sport and race vehicles), 550–555:
 driver compartment, 550–554;
 body panel removal, 555

Accident scene protection, 99–100

Acronyms, xxiii

Advanced life support (ALS), 5

Air bags, 65–67, 210–213, 362–363, 369–370, 534–536:
 labels, 213;
 entrapment procedures, 362;
 vehicle lifting procedures, 369–370;
 bus lifting procedures, 534–536

Air chisel applications (door), 153–154, 167–170, 183–187, 192–195, 403–404, 409, 498–500:
 door opening, 153–154, 403–404;
 door removal, 167–170, 409, 498–500;
 door creation, 183–187, 192–195

Air chisel applications (roof), 234–240, 256–261, 424–428, 510–511:
 roof flapping, 234–240;
 roof removal, 234–240;
 roof cutting, 256–561;
 roof entry, 424–428, 510–511

Air chisel applications (seat), 289–291, 529–531:
 seat removal, 289–291, 529–531

Air chisel applications (side), 454–456, 517–519:
 side entry, 454–456, 517–519

Air chisels (air/pneumatic tools), 52–63, 153–154, 167–170, 183–187, 192–195, 234–240, 256–261, 289–291, 403–404, 409, 424–428, 454–456, 498–500, 510–511, 517–519, 529–531:
 how it works, 52–53;
 T cutter bit, 53–57;
 dual cutter bit, 53, 56–57;
 curved cutter bit, 53, 55–62;
 ordinary/cold chisel bit, 54, 56–57, 62;
 basic cuts, 54–58;
 bit lubrication, 58;
 bit maintenance, 58–63;
 applications, 153–154, 167–170, 183–187, 192–195, 234–240, 256–261, 289–291, 403–404, 409, 424–428, 454–456, 498–500, 510–511, 517–519, 529–531

Air cylinders (apparatus-mounted), 44

Air hoses (air/pneumatic tools), 46, 48–50:
    hose couplings, 46, 50

Air pressure gauge, 46

Air pressure regulators, 45–49:
    components, 45–49;
    inlet fitting, 45, 47;
    regulator body, 46–47;
    bonnet, 46–47;
    high-pressure gauge, 46;
    low-pressure gauge, 46;
    relief valve, 46;
    hose couplings, 46;
    diaphragm, 47;
    component features, 48–49

Air sources (air/pneumatic tools), 43–45:
    SCBA cylinder, 43–44;
    large apparatus-mounted cylinders, 44;
    specialized apparatus, 44;
    set-ups to avoid, 44–45

Air/pneumatic tool systems, 42–63, 414, 417–419:
    overview, 43;
    air sources, 43–45;
    air pressure regulators, 45–49;
    air hoses, 48–50;
    4500-psi SCBA conversion, 50–51;
    setting up system, 51;
    troubleshooting, 51–52;
    air chisels, 52–63

Apparatus features and tool storage, 86–93

Apparatus placement (scene stabilization), 99–100

Apparatus. *See* Tools, equipment, and apparatus *and individual types*

Arrival size-up, 4–5:
    resource adequacy, 4–5

Assignment of tasks (company officer), 8–11

Automobile underride (trucks and tractor trailers), 393, 467–476

Automotive-repair style spreaders, 71

B

Basic life support (BLS), 5

Battery disconnection, 211–212

Bits (chisel), 52–63:
    how it works, 52–53;
    T cutter bit, 53–57;
    dual cutter bit, 53, 56–57;
    curved cutter bit, 53, 55–62;
    ordinary/cold chisel bit, 54, 56–57, 62;
    basic cuts, 54–58;
    lubrication, 58;
    maintenance, 58–63;
    sharpening, 58–63

Blade change/attachment (whizzer/die grinder), 65

Body panel removal (monster truck), 555

Bolt cutters (steering ring relocation), 294

Bonnet (air pressure regulator), 46–47

Boots, coats, and pants, 22

Bottle jack, 69–70:
    operations, 69–70

Bus emergency exit procedures, 484–487:
    school bus, 484–485;
    transit bus, 486–487

Bus extrication. *See* School buses/transit buses extrication.

Bus lifting (school buses/transit buses), 534–537:
    air bags, 534–536;
    hydraulic spreaders, 537

C

Cab-forward dash roll-up (trucks and tractor trailers), 435–436:
    hydraulic tools, 435–436

Cable hoist. See Come-along.

Cable load, 34–42:
    slack, 35–37;
    tension, 40–42

Cable tension, 40–42

Cab-over dash roll (trucks and tractor trailers), 417–419:
    pneumatic tools, 417–419;
    hand tools, 417–419

Index

Cab-over dash roll-up (trucks and tractor trailers), 437–438:
hydraulic tools, 437–438

Capabear (vehicle stabilization), 130–131, 136:
vehicle on roof, 136

Case study, 19–20:
overturned tractor trailer, 19–20

Center punch (glass removal), 148–149

Chain and shackle kit, 74–75

Chain configurations, 30–32:
come-along, 30;
hydraulic spreaders, 30–32;
chain tips, 31–32

Chain construction and grading, 28

Chain set components, 28–30:
master link, 28;
hooks, 28–30;
identification tag, 30

Chain tips, 31–32

Chains, 26–32, 74–75:
construction and grading, 28;
chain set components, 28–30;
configurations, 30–32;
tips, 31–32;
chain and shackle kit, 74–75

Chocks (vehicle stabilization), 136:
vehicle on roof, 136

Coats, pants, and boots, 22

Cold chisel/ordinary bit, 54, 55–57, 62

Come-along (pedal relocation), 307–314:
across hood, 307–308;
through door, 309–310;
through rear windshield, 311–312;
over roof, 313–314

Come-along applications, 127–128, 163–165, 244–247, 274, 280–285, 295–298, 307–314, 361, 377–379, 429–434, 521–524:
vehicle stabilization, 127–128;
door widening, 163–165;
roof flapping, 244–247;
steering column relocation, 274, 295–298, 521–524;
interior spread, 280–282;
front seat relocation, 280–285;
pedal relocation, 307–314;
entrapment, 361;
vehicle lifting, 377–379;
dash roll-up, 429–434

Come-along basic actions, 34–42:
letting out cable without load, 34–35;
taking up slack cable, 35–37;
operating with no slack and under load, 37–39;
releasing cable tension under load, 40–42

Come-along chains, 30:
configurations, 30

Come-alongs, 30, 32–42, 37–39, 127–128, 163–165, 244–247, 274, 280–285, 295–298, 307–314, 361, 377–379, 429–434, 521–524:
chain configurations, 30;
components, 32–42;
basic actions, 34–42;
applications, 127–128, 163–165, 244–247, 274, 280–285, 295–298, 307–314, 361, 377–379, 429–434, 521–524

Company officer role/responsibility (organization), 3–17:
decisions, 3;
size-up, 3–5;
hazards presence, 5;
resources, 5–8;
tactical plan development/task assignment, 8–11;
tactical options, 11–13;
risk/benefit analysis, 13–14;
plan A, B, and C, 14–15;
extrication time estimates, 15–16;
should not be doing, 16–17

Compliance issue (PPE), 24

Components (air pressure regulator), 45–49:
inlet fitting, 45, 47;
body, 46–47;
bonnet, 46–47;
pressure gauge, 46;
relief valve, 46;
hose couplings, 46;
diaphragm, 47;
features, 48–49

Components (chain set), 28–30:
master link, 28;
hooks, 28–30;
identification tag, 30

Components (come-along), 30, 32–42:
  chain configuration, 30;
  basic actions, 34–42;
  no slack and under load, 37–39

Compressed gas cylinders (roof procedures), 210, 212, 218–221

Compression stroke, 78

Construction basics (sport and race vehicles), 540–542:
  roll bars and cages, 540;
  fuel and fuel tanks, 540–541;
  steering wheels, 541;
  seats and belts, 541;
  doors and other interior access, 541–542;
  shutting down engine, 542

Conventional cab rear entry (trucks and tractor trailers), 457–458:
  reciprocating saw, 457–458

Conventional cab roll-up (trucks and tractor trailers), 439–445:
  hydraulic tools, 439–445

Crankshaft, 77

Crash-related impalement extrication, 339–353:
  procedures, 345–353

Crew safety precautions, 211–212

Cribbing, 24–27, 122–123, 136, 367:
  vehicle stabilization, 122–123, 136;
  vehicle on roof, 136;
  vehicle lifting, 367

Crushed roof lifting, 222–225:
  Hi-Lift jack, 222;
  hydraulic spreader, 223;
  hydraulic ram, 224–225

Curved cutter bit, 53, 55–62

Cutter bits, 53–62:
  T cutter bit, 53–57;
  dual cutter bit, 53, 56–57;
  curved cutter bit, 53, 55–62

Cutters. *See* Hydraulic cutters *and* Manually-operated cutters.

Cutting pipes (impalement), 349–351:
  reciprocating saw, 349;
  whizzer, 350;
  hydraulic cutters, 351

Cutting rebar (impalement), 346, 348:
  hydraulic cutters, 346;
  reciprocating saw, 348

Cutting square stock (impalement), 352:
  hydraulic cutters, 352

Cutting wood (impalement), 353:
  reciprocating saw, 353

D

Dash lift (trucks and tractor trailers), 446–448:
  cab-over, 446–448;
  hydraulic tools, 446–448

Dash lift, 332–337, 446–448:
  interior procedures, 332–337;
  trucks and tractor trailers, 446–448

Dash removal (trucks and tractor trailers), 449–451:
  cab-over, 449–451;
  reciprocating saw, 449–451

Dash roll (interior procedures), 325:
  Hi-Lift jack, 325

Dash roll (trucks and tractor trailers), 417–419:
  cab-over, 417–419;
  pneumatic tools, 417–419;
  hand tools, 417–419

Dash roll, 325, 417–419:
  interior procedures, 325;
  trucks and tractor trailers, 417–419

Dash roll-up (interior procedures), 274–275, 326–331:
  hydraulic tools, 326–331

Dash roll-up (trucks and tractor trailers), 429–445:
  reciprocating saw, 429–434;
  come-along, 429–434;
  cab-forward, 435–436;
  cab-over, 437–438;
  conventional cab, 439–445

Dash roll-up, 274–275, 326–331, 429–445:
  interior procedures, 274–275, 326–331;
  trucks and tractor trailers, 429–445

Decisions (company officer), 3

Index

Deployment (accident scene), 97, 99–100:
  personnel, 97, 99;
  equipment, 97, 99;
  lighting, 99;
  hose, 99;
  leaking fluids, 99;
  traffic, 99–100

Diaphragm (air pressure regulator), 47

Die grinder/whizzer, 63–65, 315, 350:
  operating, 63–65, 315, 350;
  blade change/attachment, 65;
  pedal removal, 315;
  cutting pipes, 350

Diesel engine, 77

Dispatch size-up, 3–4:
  access, 4;
  victim information, 4;
  terrain and geography, 4

Door and side extrication, 141–208:
  terminology, 141–142;
  size-up and action plan, 142–143;
  procedures and tools selection, 144;
  procedure steps, 144–145;
  procedures, 147–208

Door creation, 183–195, 203–205:
  third door, 183–191;
  air chisel, 183–187, 192–195;
  hydraulic tools, 188–191, 203–205;
  fourth door, 192–195, 203–205

Door entry (school buses/transit buses), 490–497:
  hand tools, 490–491;
  reciprocating saw, 492–493;
  hydraulic cutter, 494–497

Door opening (school buses/transit buses), 501:
  hydraulic spreaders, 501

Door opening (trucks and tractor trailers), 388–390, 403–408:
  hydraulic spreaders, 388–390, 405–408;
  air chisel, 403–404

Door opening procedures, 153–165, 388–390, 403–408, 501:
  air chisel, 153–154;
  hydraulic spreaders, 155–161;
  vertical spread, 157–159;
  rear door, 160–161;
  manual widening, 162;
  come-along widening, 163–165;
  trucks and tractor trailers, 388–390, 403–408;
  school buses/transit buses, 501

Door removal (school buses/transit buses), 498–500:
  air chisel, 498–500

Door removal (trucks and tractor trailers), 409–413:
  air chisel, 409;
  hydraulic tools, 410–413

Door removal procedures, 166–182, 409–413, 498–500:
  impact wrench, 166;
  air chisel, 167–170;
  spreaders, 171–176;
  cutters, 177–180;
  van door, 181–182;
  trucks and tractor trailers, 409–413;
  school buses/transit buses, 498–500

Door spreading, 155–161:
  horizontal, 155–156;
  vertical, 157–159

Door widening, 162–165:
  manual, 162;
  come-along, 163–165

Doors/access to interior (sport and race vehicles), 541–542

Drag racing doors (sport and race vehicles), 547

Dragster (sport and race vehicles), 551:
  access means, 551;
  driver compartment, 551

Driver compartment (sport and race vehicles), 550–554:
  access means, 550–554;
  funny car, 551–552;
  dragster, 551;
  monster truck, 553–554

Driver/operator role/responsibility, 17–18

Driver's seat back removal (school buses/transit buses), 525–527:
  hydraulic tools, 525–527

Dual cutter bit, 53, 56–57

Dump/selector valve, 74

E

Electric motor, 77

Electric power distribution, 90

Electric reciprocating saw. SEE Reciprocating saw.

Electric tools, 67–68

Electrical hazard, 101–103:
live wires, 103

Electrical shutdown switches (sport and race vehicles), 548

Emergency door opening (school buses/transit buses), 501

Emergency exit (school buses/transit buses), 478–479, 481, 484–487, 498–500:
school bus, 484–485;
transit bus, 486–487;
exit door removal, 498–500

Emergency exit door removal (school buses/transit buses), 498–500:
air chisel, 498–500

Emergency exit roof hatch (school buses/transit buses), 485–486:
school bus, 485;
transit bus, 486

Emergency medical technician (EMT), 6–7

Engine compartment access, 114–121:
Halligan adz, 114;
Halligan pike, 115–117;
saw, 118–121

Engine lubrication (power plants), 80

Engine shutdown, 103, 105, 211, 542, 548:
sport and race vehicles, 542, 548;
shutdown switches, 548

Engine-driven hydraulic tools, 72–85:
history, 72–76;
rescue tool system, 72–73;
modern spreaders, cutters, and rams, 76–77;
power plants, 77–85

Entrapment beneath/between vehicles (extrication), 355–383:
general procedures, 357–363;
hand tools and wedges, 360;
pry bars, 360;
Hi-Lift jack, 360–361;
come-along, 361;
manually-operated spreaders, 361–362;
high-pressure air bags, 362;
heavy hydraulic spreaders, 363;
procedures, 365–383

Equipment deployment, 97, 99

Equipment. *See* Tools, equipment, and apparatus *and individual types.*

Estimated time of arrival (ETA), 15–16

Exhaust stroke, 79

Exit door removal (school buses/transit buses), 498–500:
air chisel, 498–500

Extrication time estimates, 15–16

Eye, face, and head protection, 23–24

F

Face, eye, and head protection, 23–24

Field surgical kit/amputation protocol, 7

Fire control/extinguishment, 107–108

Firefighter role/responsibility, 18

Firefighters, tools, and leadership, 1–20:
organization, 1–2

Floor jack, 70–71:
operations, 70–71

Forward movement (vehicle stabilization), 124–126

Four-stroke engine, 77–80

Fourth door creation, 192–195, 203–205:
air chisel, 192–195;
hydraulic tools, 203–205

Front door entry (school buses/transit buses), 490–497:
hand tools, 490–491;
reciprocating saw, 492–493;
hydraulic cutter, 494–497

Front seat relocation (interior procedures), 283–285:
come-along, 283–285

Index

Front seat removal (interior procedures), 289–291, 293:
  air chisel, 289–291;
  hydraulic tools, 293

Fuel (power plants), 80–81:
  fuel choice, 81

Fuel and fuel tanks (sport and race vehicles), 540–541

Fuel selection, 81, 83

Funny car (sport and race vehicles), 550–552:
  access means, 550–551;
  driver compartment, 551–552

G

Gas cylinders (roof procedures), 210, 212, 218–221

Gas lifters (roof), 210, 214–215

Glass removal (school buses/transit buses), 488–489:
  hand tools, 488–489

Glass removal, 148–150, 226–229, 401–402, 480, 488–489:
  center punch, 148–149;
  Halligan, 150;
  hand tools, 401–402, 488–489;
  school buses/transit buses, 488–489

Gloves, 22–23

H

Halligan (engine compartment access), 114–117:
  Halligan adz, 114;
  Halligan pike, 115–117

Halligan applications, 114–117, 150–151:
  engine compartment access, 114–117;
  glass removal, 150;
  purchase point, 151

Hand tools (glass removal), 401–402, 488–489:
  windshield glass, 401–402, 488;
  side glass, 489

Hand tools applications, 206–207, 231–233, 360, 366, 401–402, 417–419, 488–491:
  trunk opening, 206–207;
  roof flapping, 231–233;
  entrapment, 360, 366;
  windshield removal, 401–402, 488;
  glass removal, 401–402, 488–489;
  dash roll, 417–419;
  front door entry, 490–491

Hand tools, 93, 206–207, 231–233, 360, 366, 401–402, 417–419, 488–491:
  applications, 206–207, 231–233, 360, 366, 401–402, 417–419, 488–491

Hard surface vehicle lifting (entrapment), 373–374:
  Hi-Lift jack, 373–374

Hazard recognition, 5, 101–103:
  traffic, 5;
  utility service damage, 5;
  vehicles involved, 5;
  electrical hazard, 101–103;
  live wires, 103

Hazards presence (company officer), 5

Head, face, and eye protection, 23–24

Heat-Fence (crash-related impalement), 347

Heavy duty recovery vehicle, 9–11

Heavy hydraulic spreaders, 363

Helicopter landing zone (HLZ), 6

High-pressure air bags (lifting), 362–363, 369–370

High-pressure gauge (air pressure regulator), 46

Hi-Lift jack (vehicle lifting), 373–376, 380–381:
  hard surface, 373–374;
  soft surface, 375–376

Hi-Lift jack applications, 222, 276, 319–320, 325, 360–361, 373–376, 380–381:
  roof lifting, 222;
  interior space creation, 276, 319–320;
  dash roll, 325;
  entrapment, 360–361;
  vehicle lifting, 373–376, 380–381

Hi-Lift jack, 222, 276, 319–320, 325, 360–361, 373–376, 380–381:
  applications, 222, 276, 319–320, 325, 360–361, 373–376, 380–381

Hood opening/removal (trucks and tractor trailers), 396–397

Hooks (chain), 28–30

Horizontal spreading (door), 155–156

Hose couplings (air pressure regulator), 46, 50

Hose deployment (firefighting), 99

Hose reel (hydraulic system), 74–76

Hoses (air/pneumatic tools), 46, 48–50:
couplings, 46, 50

Hospital/trauma center assistance, 6–7

Hybrid vehicles, 105–106

Hydraulic cutters (door removal), 177–180:
interior approach, 177–178;
exterior approach, 179–180

Hydraulic cutters (pedal removal), 316–318:
manually-operated, 316–317

Hydraulic cutters applications, 69, 76–77, 177–180, 215, 269–271, 292, 305, 316–318, 346, 351–352, 416, 464–466, 494–497, 504, 520:
door removal, 177–180;
roof flapping, 269–271;
steering wheel removal, 305, 416;
pedal removal, 316–318;
rebar, 346;
large pipes, 351;
square stock, 352;
rear entry, 464–466;
front door entry, 494–497;
window posts removal, 504;
steering ring removal, 520

Hydraulic cutters, 69, 76–77, 177–180, 215, 269–271, 292, 305, 316–318, 346, 351–352, 416, 464–466, 494–497, 504, 520:
manually-operated, 69, 316–317;
applications, 69, 76–77, 177–180, 215, 269–271, 292, 305, 316–318, 346, 351–352, 416, 464–466, 494–497, 504, 520
seat back removal, 292

Hydraulic fluid, 74, 81–84:
fluid reservoir, 84

Hydraulic ram applications, 224–225, 321–322:
roof lifting, 224–225;
interior spread, 321–322

Hydraulic rams, 76–77, 224–225, 321–322:
applications, 224–225, 321–322

Hydraulic reservoir (power plants), 81–82

Hydraulic spreaders (door opening), 155–161, 388–390, 405–408, 501:
horizontal, 155–156;
vertical, 157–159;
rear door, 160–161;
emergency door, 501

Hydraulic spreaders (door removal), 171–176:
exterior approach, 171–174;
interior approach, 175–176

Hydraulic spreaders applications, 155–161, 208, 223, 274, 288, 299–304, 363, 371–372, 388–390, 405–408, 420, 461–463, 501, 537:
door opening, 155–161, 388–390, 405–408, 501;
trunk opening, 208;
roof lifting, 223;
steering column relocation, 274, 299–304;
seat movement, 288;
entrapment, 363;
vehicle lifting, 371–372;
roof tenting, 420;
rear entry, 461–463;
emergency door opening, 501;
bus lifting, 537

Hydraulic spreaders chains, 30–32:
configurations, 30–32

Hydraulic spreaders, 30–32, 68–69, 71–72, 74–77, 155–161, 208, 223, 274, 299–304, 361–363, 371–372, 388–390, 405–408, 420, 461–463, 501, 537:
chains, 30–32;
manually-operated, 69, 71–72, 361–362;
automotive-repair style, 71–72;
modern spreader tools, 71–72;
operations, 72;
accessory kit, 74–75;
applications, 155–161, 208, 223, 274, 299–304, 363, 371–372, 388–390, 405–408, 420, 461–463, 501, 537

Hydraulic tools (dash roll-up), 274–275, 326–331, 435–445:
cab-forward, 435–436;
cab-over, 437–438;
conventional cab, 439–445

Hydraulic tools (door creation), 188–191, 203–205:
third door, 188–191;
fourth door, 203–205

Index

Hydraulic tools applications, 69–72, 87, 188–191, 203–205, 241–243, 251–255, 274–275, 293, 326–331, 368, 382–383, 410–413, 415, 435–448, 505–507, 525–527, 532–533:
storage, 87;
door creation, 188–191, 203–205;
roof flapping, 241–243;
roof removal, 251–255;
dash roll-up, 274–275, 326–331, 435–445;
seat removal, 293, 532–533;
front seat removal, 293;
vehicle lifting, 368;
vehicle spreading, 382–383;
door removal, 410–413, 415;
cab-over dash lift, 446–448;
rear wall removal, 505–507;
seat back removal, 525–527

Hydraulic tools, 69–85, 87, 188–191, 203–205, 241–243, 251–255, 274–275, 293, 326–331, 368, 382–383, 410–413, 415, 435–448, 505–507, 525–527, 532–533:
manually-operated, 69–72, 368;
engine-driven, 72–85;
applications, 69–72, 87, 188–191, 203–205, 241–243, 251–255, 274–275, 293, 326–331, 368, 382–383, 410–413, 415, 435–448, 505–507, 525–527, 532–533

I

Identification tag (chain), 30

Impact wrench, 63, 68, 166:
door removal, 166

Impalement (crash-related), 339–353:
extrication, 339–353;
procedures, 345–353

Inflatable curtain, 210, 213, 218–221:
roof, 218–221

Inlet fitting (air pressure regulator), 45, 47

Intake stroke, 77–78

Interior extrication, 273–337:
steering column relocation, 274;
dash roll-up, 274–275;
pedal removal, 275–276;
pedal relocation, 276;
interior space creation, 276;
seat relocation, 276;
summary, 277;
procedures, 279–337

Interior obstructions (sport and race vehicles), 549

Interior space creation (interior procedures), 276, 319–320:
Hi-Lift jack, 276, 319–320

Interior spread procedures, 280–282, 286–287, 321–322:
come-along, 280–282;
winch, 286–287;
hydraulic ram, 321–322

J

Jacks, 69–71, 138–139:
bottle jack, 69–70;
floor jack, 70–71;
stump screw jack, 138–139

Jaws of life, 72

Jersey barrier (vehicle stabilization), 140

K

Knowledge and skills, 2

L

Laminated glass removal, 226–229

Leaking fluids, 99

Letting out cable without load (come-along), 34–35

Life support, 5

Light/lighting apparatus, 91–92, 99:
deployment, 99

Live wires contact (scene stabilization), 103

Low-pressure gauge (air pressure regulator), 46

Lubrication (bits), 58

M–N

Maintenance (bits), 58–63:
   sharpening, 58–63

Manifold block, 74, 76

Manually-operated cutters, 69, 316–317:
   pedal removal, 316–317

Manually-operated hydraulic tools, 69–72, 368:
   bottle jacks, 69–70;
   floor jacks, 70–71;
   spreaders, 71–72;
   vehicle lifting, 368

Manually-operated spreaders, 69, 71–72, 361–362:
   entrapment, 361–362

Mass casualty incident (MCI), 5–6

Master link (chain), 28

Medical units, 5–6, 340

Monster truck (sport and race vehicles), 553–555:
   access means, 553–555;
   driver compartment, 553–554;
   body panels removal, 555

O

Obstructions (sport and race vehicles), 549

Occupant/victim stabilization, 110–111

One-side stabilization (vehicle), 132–135

Ordinary/cold chisel bit, 54, 56–57, 62

Organization, 1–20:
   firefighters, tools, and leadership, 1–2;
   knowledge and skills, 2;
   company officer role/responsibility, 3–17;
   driver operator role/responsibility, 17–18;
   firefighter role/responsibility, 18;
   summary, 18;
   resource utilization, 19–20

Overturned tractor trailer (case study), 19–20

Overturned vehicle lifting (entrapment), 380–381:
   Hi-Lift jack, 380–381

P–Q

Pants, coats, and boots, 22

Paramedics, 6–7

Pedal relocation (interior procedures), 306–314:
   rope, 306;
   come-along, 307–314

Pedal relocation, 276, 306–314:
   rope, 276:
   interior procedures, 306–314

Pedal removal (interior procedures), 315–318:
   whizzer, 315;
   hydraulic cutters, 316–318

Pedal removal (trucks and tractor trailers), 414–415:
   pneumatic tools, 414;
   hydraulic tools, 415

Pedal removal, 275–276, 315–318, 414–415:
   interior procedures, 315–318;
   trucks and tractor trailers, 414–415

Personal protective equipment (PPE), 21–24, 76:
   coats, pants, and boots, 22;
   gloves, 22–23;
   head, face, and eye protection, 23–24;
   compliance issue, 24

Personnel deployment, 97, 99

Personnel safety precautions, 210–212

Pinch and curl (purchase point), 152

Pipe cutting (impalement), 349–351:
   reciprocating saw, 349;
   whizzer, 350;
   hydraulic cutters, 351

Piston, 77

Plan development (company officer), 8–11

Planning, 8–11, 14–15:
   plan development, 8–11;
   plans, 14–15

Pneumatic tools, 414, 417–419:
   door removal, 414;
   dash roll, 417–419

Pneumatic/air tool systems, 42–63, 414, 417–419:
   overview, 43;
   air sources, 43–45;
   air pressure regulators, 45–49;

Index

air hoses, 48–50;
4500-psi SCBA conversion, 50–51;
setting up system, 51;
troubleshooting, 51–52;
air chisels, 52–63

Police cars (interior procedures), 323–324:
problems, 323–324

Popping procedures (opening/removal), 153–161, 181–182, 206–208, 388–390, 396–397, 401–408, 479, 488, 501:
door: 153–161, 181–182, 388–390, 403–408, 501;
trunk, 206–208;
hood, 396–397;
windshield, 401–402, 479, 488

Power plants (hydraulic tools), 77–85:
power source, 77–80;
engine lubrication, 80;
fuel, 80–81;
pump and hydraulic reservoir, 81–82;
starting, 82–84;
shutting down, 84–85;
returning to service, 85

Power source (power plants), 77–80

Power stroke, 78

Power take off (PTO) generator, 77, 86

Pre-event training (sport and race vehicles), 546

Problem solution, 14–15

Procedures (crash-related impalement), 345–353:
cutting rebar, 346, 348;
Heat-Fence, 347;
cutting large pipes, 349–351;
cutting square stock, 352;
cutting wood, 353

Procedures (door and side), 147–208:
tempered glass removal, 148–150;
purchase point creation, 151–152;
door opening, 153–165;
door removal, 166–180;
van door opening, 181–182;
third door creation, 183–191;
fourth door creation, 192–195, 203–205;
side-out procedure, 196–202;
trunk opening, 206–208

Procedures (entrapment beneath/between vehicles), 365–382:
vehicle lifting, 366–381;
vehicle spreading, 382–383

Procedures (interior), 279–337:
interior spread, 280–282, 286–287, 321–322;
front seat relocation, 283–285;
seat movement rearward, 288;
front seat removal, 289–291, 293;
seat back removal, 292;
steering ring relocation, 294;
steering column relocation, 295–304;
steering wheel removal, 305;
pedal relocation, 306–314;
pedal removal, 315–318;
interior space creation, 319–320;
special problems in police cars, 323–324;
dash roll, 325;
dash roll-up, 326–331;
dash lift, 332–337

Procedures (roof), 217–271:
inflatable curtains, 218–221;
compressed gas cylinders, 218–221;
crushed roof lifting, 222–225;
laminated glass removal, 226–229;
roof flapping, 230–247, 269–271;
roof removal, 248–255;
three-sided roof cut, 256–266;
roof center removal, 267–268

Procedures (school buses/transit buses), 483–537:
emergency exit, 484–487;
windshield glass removal, 488;
side glass removal, 489;
front door entry, 490–497;
emergency exit door removal, 498–500;
emergency door opening, 501;
rear window posts removal, 502–504;
rear wall removal, 505–507;
roof entry, 508–511;
side wall removal, 512–516;
side entry, 517–519;
steering ring removal, 520;
steering column relocation, 521–524;
driver's seat back removal, 525–527;
seat post removal, 528;
seat removal, 529–533;
lifting bus, 534–537

Procedures (sport and race vehicles), 545–555:
  pre-event training, 546;
  drag racing doors, 547;
  remote electrical shutdown switches, 548;
  interior obstructions, 549;
  gaining access, 550;
  driver compartments, 551–554;
  body panel removal, 555

Procedures (stabilization), 113–140:
  engine compartment access, 114–121;
  vehicle cribbing, 122–123, 136;
  forward movement, 124–126;
  come-along, 127–128;
  4x4 post, 129;
  Capabear, 130–131, 136;
  one side stabilization, 132–135;
  vehicle on roof, 136–137;
  chocks, 136;
  stump screw jacks, 138–139;
  Jersey barrier, 140

Procedures (trucks and tractor trailers), 395–476:
  hood opening and removing, 396–397;
  truck stabilization, 398–400;
  windshield removal, 401–402;
  door opening, 403–408;
  door removal, 409–413;
  pedal removal, 414–415;
  steering wheel removal, 416;
  dash roll, 417–419;
  roof tenting, 420;
  roof entry, 421–428;
  dash roll-up, 429–445;
  dash lift, 446–448;
  dash removal, 449–451;
  sleeper side entry, 452–456;
  rear entry, 457–458;
  sleeper rear entry, 459–466;
  trailer underride, 467–476

Protecting accident scene, 99–100

Pry bar, 360, 367:
  entrapment, 360;
  vehicle lifting, 367

Pump and hydraulic reservoir (power plants), 81–82

Purchase point creation, 151–152:
  Halligan, 151;
  pinch and curl, 152

R

Race and sport vehicles extrication, 539–555:
  construction basics, 540–542;
  summary, 542;
  considerations, 545–555

Ram accessory kit, 74–75

Rams, 74–77, 224–225, 321–322:
  accessory kit, 74–75;
  roof lifting, 224–225;
  interior spread, 321–322

Rear door spreading, 160–161

Rear entry (sleeper), 459–466:
  reciprocating saw, 459–460;
  hydraulic spreaders, 461–463;
  hydraulic cutters, 464–466

Rear entry (trucks and tractor trailers), 457–466:
  conventional cab, 457–458;
  reciprocating saw, 457–458;
  sleeper, 459–466

Rear wall removal (school buses/transit buses), 505–507:
  hydraulic tools, 505–507

Rear window posts removal (school buses/transit buses), 502–504:
  reciprocating saw, 502–503;
  hydraulic cutter, 504

Rebar cutting (impalement), 346, 348:
  hydraulic cutters, 346;
  reciprocating saw, 348

Reciprocating saw applications, 118–121, 248–250, 262–268, 348–349, 353, 421–423, 429–434, 449–453, 457–460, 492–493, 502–503, 508–509, 512–516, 528:
  engine compartment access, 118–121;
  roof removal, 248–250;
  roof cutting, 262–266;
  roof center removal, 267–268;
  cutting rebar, 348;
  cutting large pipes, 349;
  cutting wood, 353;
  roof entry, 421–423, 508–509;
  dash roll-up, 429–434;
  cab-over dash removal, 449–451;
  sleeper side entry, 452–453;

Index

conventional cab rear entry, 457–458;
rear entry, 459–460;
front door entry, 492–493;
window posts removal, 502–503;
side wall removal, 512–516;
seat post removal, 528

Reciprocating saws, 67–68, 118–121, 248–250, 262–268, 348–349, 353, 421–423, 429–434, 449–453, 457–460, 492–493, 502–503, 508–509, 512–516, 528:
applications, 118–121, 248–250, 262–268, 348–349, 353, 421–423, 429–434, 449–453, 457–460, 492–493, 502–503, 508–509, 512–516, 528

Recovery vehicle, 9–11

Regulator body (air pressure), 46–47

Regulators (air pressure), 45–49:
inlet fitting, 45, 47;
regulator body, 46–47;
bonnet, 46–47;
high-pressure gauge, 46;
low-pressure gauge, 46;
relief valve, 46;
hose couplings, 46;
diaphragm, 47;
components and features, 48–49

Releasing cable tension under load (come-along), 40–42

Relief valve (air pressure regulator), 46

Remote electrical shutdown switches (sport and race vehicles), 548

Rescue tool system, 72–73

Resource utilization, 19–20:
case study, 19–20

Resources determination (company officer), 5–8:
medical units, 5–6;
hospital/trauma center assistance, 6–7;
suppression units, 6;
specialty units, 8;
wreckers, 8

Returning to service (power plants), 85

Risk/benefit analysis, 13–14

Roadway hazards, 341

Roll bars and cages (sport and race vehicles), 540

Roof center removal, 267–268:
reciprocating saw, 267–268

Roof cutting, 256–266:
air chisel, 256–261;
reciprocating saw, 262–266

Roof entry (school buses/transit buses), 508–511:
reciprocating saw, 508–509;
air chisel, 510–511

Roof entry (trucks and tractor trailers), 421–428:
reciprocating saw, 421–423;
air chisel, 424–428

Roof entry, 421–428, 508–511:
trucks and tractor trailers, 421–428;
school buses/transit buses, 508–511

Roof extrication, 209–271:
roof structure, 210;
size-up of roof, 210;
crew safety precautions, 211–212;
air bag labels, 213;
technology advancements, 213–214;
gas lifters, 214–215;
tools selection, 215;
procedure selection, 216;
summary, 216;
procedures, 217–271

Roof flapping (vehicle on side), 269–271:
hydraulic cutters, 269–271

Roof flapping, 230–247, 269–271:
striking, 230;
hand tools, 231–233;
air chisel, 234–240;
hydraulic tools, 241–243;
come-along, 244–247;
vehicle on side, 269–271

Roof hatch emergency exit (school buses/transit buses), 485–486:
school bus, 485;
transit bus, 486

Roof lifting, 222–225:
Hi-Lift jack, 222;
hydraulic spreader, 223;
hydraulic ram, 224–225

Roof removal, 230–255:
air chisel, 234–240;
reciprocating saw, 248–250;
hydraulic tools, 251–255

Roof size-up, 210

Roof structure, 210

Roof tenting (trucks and tractor trailers), 420:
 hydraulic spreaders, 420

Rope (pedal relocation), 276, 306

S

Safety precautions (crew), 210–212

Saw. SEE Reciprocating saws.

Scene lighting, 91–92

Scene stabilization, 95–103:
 protection, 99–100;
 apparatus placement, 99–100;
 utility hazards, 101–103;
 live wires contact, 103

School bus emergency exit, 484–485:
 side door, 484;
 roof hatch, 485

School buses/transit buses extrication, 477–537:
 procedures, 483–537

Screw jack (vehicle stabilization), 138–139:
 vehicle on roof, 138–139

Seat back removal, 292, 525–527:
 hydraulic shears, 292;
 school buses/transit buses, 525–527

Seat movement rearward (interior procedures), 288:
 spreaders, 288

Seat post removal (school buses/transit buses), 528:
 reciprocating saw, 528

Seat relocation (interior procedures) 276, 280–285:
 come-along, 280–285

Seat removal (interior procedures), 289–291, 293:
 air chisel, 289–291;
 hydraulic tools, 293

Seat removal (school buses/transit buses), 529–533:
 air chisel, 529–531;
 hydraulic tools, 532–533

Seats and belts (sport and race vehicles), 541

Self-contained breathing apparatus (SCBA), 43–45, 50–51:
 SCBA cylinder, 43–44;
 large apparatus-mounted cylinders, 44;
 specialized apparatus, 44;
 set-ups to avoid, 44–45;
 conversion to 4500-psi SCBAs, 50–51

Set-ups to avoid (air sources), 44–45

Sharpening (bits), 58–63

Shears (seat back removal), 292

Shutdown switches (sport and race vehicles), 548

Shutting down (power plants), 84–85

Shutting down vehicle, 103–105, 211, 542, 548:
 shutdown switches, 548

Side door emergency exit (school buses), 484

Side entry (school buses/transit buses), 517–519:
 air chisel, 517–519

Side entry (trucks and tractor trailers sleeper), 452–456:
 reciprocating saw, 452–453;
 air chisel, 454–456

Side glass removal (school buses/transit buses), 489

Side wall removal (school buses/transit buses), 512–516:
 reciprocating saw, 512–516

Side-impact protection system (SIPS), 210, 213

Side-out procedures, 196–202

Size-up, 3–5, 142–143, 210, 339–342:
 dispatch, 3–4;
 upon arrival, 4–5;
 door and side, 142–143;
 roof, 210;
 impalement, 339–342

Skills and knowledge, 2

Slack cable take-up (come-along), 35–37

Sleeper rear entry (trucks and tractor trailers), 459–466:
 reciprocating saw, 459–460;
 hydraulic spreaders, 461–463;
 hydraulic cutters, 464–466

Sleeper side entry (trucks and tractor trailers), 452–456:
 reciprocating saw, 452–453;
 air chisel, 454–456

Index

Sleepers, 390–392, 452–456, 459–466:
 side entry, 452–456;
 rear entry, 459–466

Soft soil vehicle lifting (entrapment), 377–379:
 come-along, 377–379

Soft surface vehicle lifting (entrapment), 375–376:
 Hi-Lift jack, 375–376

Space creation (interior procedures), 276, 319–320:
 Hi-Lift jack, 276, 319–320

Specialized air source apparatus, 44

Specialty units, 8

Sport and race vehicles extrication, 539–555:
 construction basics, 540–542;
 summary, 542;
 considerations, 545–555

Spreaders. SEE Hydraulic spreaders AND Manually-operated spreaders.

Square stack cutting (impalement), 352:
 hydraulic cutters, 352

Stabilization (trucks and tractor trailers), 398–400

Stabilization procedures (scene/vehicle/occupants), 95–140, 398–400:
 scene stabilization, 95–103;
 vehicle stabilization, 103–110;
 victim/occupant stabilization, 110–111;
 summary, 111;
 procedures, 113–140

Starting (power plants), 82–84

Steering column relocation (interior procedures), 295–304:
 come-along, 295–298;
 hydraulic spreader, 299–304

Steering column relocation (school buses/transit buses), 521–524:
 come-along, 521–524

Steering column relocation, 274, 295–304, 521–524:
 hydraulic spreader, 274;
 come-along, 274;
 interior procedures, 295–304;
 school buses/transit buses, 521–524

Steering ring relocation, 294:
 bolt cutters, 294

Steering ring removal (school buses/transit buses), 520:
 hydraulic cutters, 520

Steering wheel removal, 305, 416:
 hydraulic cutters, 305, 416:
 trucks and tractor trailers, 416

Steering wheels (sport and race vehicles), 541

Step chocks (vehicle on roof), 136

Striking roof procedures, 230

Stump screw jack (vehicle stabilization), 138–139:
 vehicle on roof, 138–139

Supplemental restraint system, 213

Suppression units, 6, 17–18

System set-up (air/pneumatic tools), 51

T

T cutter bit, 53–57

Tactical options, 11–13:
 spreading, 12;
 pulling, 12;
 removing, 12–13;
 securing, 12

Tactical plan development (company officer), 8–13:
 options, 11–13

Taking up slack cable (come-along), 35–37

Task assignment (company officer), 8–11

Technology advancements (roof), 213–214

Tempered glass removal, 148–150:
 center punch, 148–149;
 Halligan, 150

Terminology, xxiii, 141–142, 385–388:
 acronyms, xxiii;
 door and side procedures, 141–142;
 trucks and tractor trailers, 385–388

Third door creation, 183–191:
 air chisel, 183–187;
 hydraulic tools, 188–191

Three-sided roof cutting, 256–266:
 air chisel, 256–261;
 reciprocating saw, 262–266

Time estimates (extrication), 15–16

Tool storage, 86–93

Tools and equipment descriptions, 24–85:
cribbing, 24–25;
chains, 26–32;
come-alongs, 32–42;
air/pneumatic tool systems, 42–63;
impact wrenches, 63;
whizzer/die grinder, 63–65;
air bags, 65–67;
electric tools, 67–68;
manual hydraulic tools, 69–72;
engine-driven hydraulic tools, 72–85

Tools selection (roof procedures), 215

Tools, equipment, and apparatus, 21–93, 99–100:
introduction, 21;
personal protective equipment, 21–24;
tools and equipment, 24–85;
apparatus, 85, 99–100;
apparatus features and tool storage, 86–93

Tractor trailer overturn, 19–20:
case study, 19–20

Tractor trailers and trucks extrication, 19–20, 385–476:
case study, 19–20;
terminology, 385–388;
opening door, 388–390;
dash roll-up, 390–392;
trailer underride by automobile, 393;
summary, 393;
procedures, 395–476

Traffic management, 99–100

Trailer underride (trucks and tractor trailers), 393, 467–476:
automobile, 393, 467–476

Training event (sport and race vehicles), 546

Transit bus emergency exit, 486–487:
roof hatch, 486;
window, 487

Transit buses/school buses extrication, 477–537:
procedures, 483–537

Trauma center/hospital assistance, 6–7

Troubleshooting (air/pneumatic tools), 51–52

Trucks and tractor trailers extrication, 19–20, 385–476:
case study, 19–20;
terminology, 385–388;
door opening, 388–390;
dash roll-up, 390–392;
trailer underride by automobile, 393;
summary, 393;
procedures, 395–476

Trunk opening procedures, 206–208:
hand tools, 206–207;
hydraulic spreaders, 208

Two-side stabilization (vehicle), 130–131

Two-stroke engine, 77

U

Underride (trucks and tractor trailers), 467–476

Universal precautions, 210–212

Upon arrival, 4–5:
size-up, 4–5

Utility hazards (scene stabilization), 101–103

V

Van door opening/removal, 181–182

Vehicle components, 141–142

Vehicle lifting (entrapment), 355–363, 366–381:
wedges, 366;
pry bar and cribbing, 367;
manual hydraulic tools, 368;
air bags, 369–370;
hydraulic spreaders, 371–372;
hard surface, 373–374;
soft surface, 375–376;
soft soil, 377–379;
overturned vehicle, 380–381

Vehicle movement, 108–110, 124–126:
stopping movement, 108–110

Vehicle on roof (stabilization), 136–137:
step chocks, 136;
capabear, 136;
cribbing, 136;
stump screw jack, 138–139

Index

Vehicle on side, 122–123, 127–135:
cribbing, 122–123;
come-along, 127–128;
4x4 post, 129;
stabilization, 130–135

Vehicle shutdown, 103–105, 211, 542–548:
shutdown switches, 548

Vehicle spreading (entrapment), 382–383:
hydraulic tools, 382–383

Vehicle stabilization from side(s), 130–135:
capabear, 130–131

Vehicle stabilization, 103–110, 130–135:
shutting down vehicle, 103–105;
hybrid vehicles, 105–106;
fire control/extinguishment, 107–108;
stopping movement, 108–110;
from side(s), 130–135

Vertical spreading (door), 157–159

Victim/occupant stabilization, 110–111

Window posts removal (school buses/transit buses), 502–504:
reciprocating saw, 502–503;
hydraulic cutter, 504

Windshield glass removal (school buses/transit buses), 479, 488:
hand tools, 488

Windshield removal (trucks and tractor trailers), 401–402:
hand tools, 401–402

Wood cutting (impalement), 353:
reciprocating saw, 353

Wrecker assistance, 8

Wrench (impact), 63, 68, 166:
door removal, 166

W–Z

Wall removal (school buses/transit buses), 505–507, 512–516:
rear wall, 505–507;
hydraulic tools, 505–507;
side wall, 512–516;
reciprocating saw, 512–516

Warning devices, 98

Wedges (entrapment), 360, 366

Whizzer/die grinder, 63–65, 315, 350:
operating, 63–65, 315, 350;
blade change/attachment, 65;
pedal removal, 315;
cutting pipes, 350

Widening (door), 162–165:
manual, 162;
come-along, 163–165

Winch (interior spread), 286–287

Window emergency exit (transit buses), 487

The Rules of Rescue Have Changed

Approx 433 pages/Hardcover/8 1/2 x 11/August 2004
ISBN 1-59370-014-8 $89.00 US
Shipping & tax may be added

TECHNICAL RESCUE OPERATIONS, VOLUME I: PLANNING, TRAINING, AND COMMAND
by Larry Collins, Captain, Los Angeles Fire Department

1st in a three-volume series!

The new rules of rescue have improved the way firefighters and other first responders are able to detect, locate, treat, and extract trapped victims in practically any situation by emphasizing factors like innovative training, better equipment, and more intelligent planning by well-informed decision-makers.

Whether you live in an urban or rural fire district, working for a paid or volunteer department, increasingly your fire department will be called upon for rescue. *Technical Rescue Operations, Volume I* is a groundbreaking new book that lays the framework for responding to disasters and technical rescues of all types in the safest and most effective manner.

Key Features & Benefits:
- Fully explains the concept and origins of rescue
- Concepts taught in this book will provide guidance for rescue teams of any size and budget
- Explains how to develop, manage, and operate a modern rescue system
- Illustrates how to apply principles of rescue to actual high-risk operations
- Includes more than 100 photos

Contents:
Introduction • Planning for "Daily" Urban Search and Technical Rescues • Planning for Disaster Search and Technical Rescue Operations • Developing Multi-Tiered USAR/Rescue Systems • Rescue Apparatus and Equipment • Rescue Training • Commanding Urban Search and Technical Rescue Operations • References • Appendix • Bibliography

About the Author:
Larry Collins has been a captain of the Los Angeles County Fire Department's central USAR Company since 1990. He has been involved in rescue operations worldwide, including the Pentagon collapse during the September 11, 2001 attacks, and the Alfred P. Murrah Building bombing in Oklahoma City, as well as numerous natural disasters.

For more information visit
www.technicalrescueops.com

More than 100 photos!

3 Easy Ways to Order:
1. Tel: 1.800.752.9764 or +1.918.831.9421
2. Online: www.FireEngineeringBooks.com
3. E-mail: sales@pennwell.com

MAXIMIZE YOUR TRAINING EFFORTS

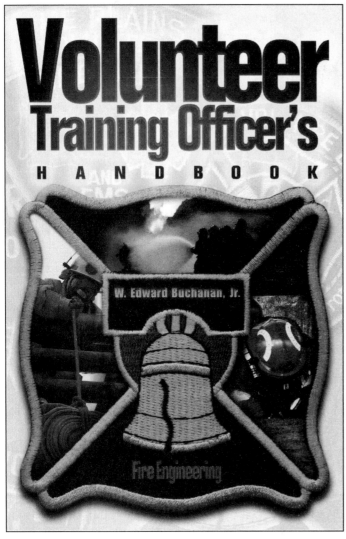

180 pages/Softcover/6x9/2003
ISBN 0-87814-834-5 $55.00 US
Shipping & tax may be added

VOLUNTEER TRAINING OFFICER'S HANDBOOK
by Eddie Buchanan, Division Chief,
Hanover County (VA) Fire Dept.

New training officers have the challenging and rewarding task of finding the proper balance between change and tradition as the fire service evolves.

This comprehensive guide, by Eddie Buchanan, will provide the volunteer training officer everything needed to develop and implement a successful program while promoting quality and discipline in today's volunteer firefighter. Alternative delivery options will be presented that allow the volunteer to achieve their training goals while meeting their responsibilities at home.

Each book comes with a CD-ROM that includes appendices and instructor materials such as roll call forms, PowerPoint presentations, and note-taking sheets for students.

Features & Benefits:
- Step-by-step implementation process for the volunteer training officer
- Scheduling templates that meet the training officer's core needs
- Core training values such as accountability, discipline, and pride

Contents:
- Self-preparation
- Assessing department needs
- A systems approach to training
- The Academy, Part 1: Laying the training foundation
- The Academy, Part 2: The key components
- In-service training options
- Comprehensive scheduling
- Alternative training options and marketing
- Biography

Best Seller!

3 Easy Ways to Order:
1. Tel: 1.800.752.9764 or +1.918.831.9421
2. Online: www.FireEngineeringBooks.com
3. E-mail: sales@pennwell.com

PennWell®